CONTENTS

FOREWORD
(First Edition)

During recent years with the rising need for world food supplies interest in rice production has increased rapidly. This interest has been heightened by the development of the new rice varieties from the International Rice Research Institute in the Philippines. These varieties have a very high yield potential; but they do need to be grown with the best possible techniques to realise this potential, which perhaps accounts for the renewed interest in improved rice agronomy, mechanised cultivation, irrigation, and above all in prevention of crop losses from pest attacks. Many research institutes and experimental stations are now carrying out trials to find the best methods of growing the new rices in their areas so that they can advise growers. This has lead to improvements in the methods of growing all types of rice. With so much happening in the field of rice growing this Manual has been written in an attempt to assemble all aspects of pest control and inform readers of the potential losses from insects, diseases, weeds, rats, etc.

It is hoped that the Manual will be useful to people directly concerned with growing rice, and especially to those concerned with any aspect of pest control. In order to avoid confusion with nomenclature the following terms have been used throughout the text:

Rough Rice or Paddy: Rice grain in the husk after threshing.

Husked Rice: Rice from which the husk only has been removed but which still retains the bran layers and most of the germs. (Also known as Brown Rice, Cargo Rice or Loonzain Rice).

Milled Rice: (Fully milled) Rice from which the husks (hulls), germ and bran layers have been removed by machinery. Also known as Polished Rice or White Rice.

Parboiled Rice: Rice which has been processed by steaming or soaking in water, heating (usually by steam) and drying before being milled. Paddy may be parboiled and called Parboiled Paddy before it is milled.

Paddy: A flooded field in which transplanted rice is grown.

During the preparation of the Manual many people have been very helpful and provided information and assistance of all kinds. The editor would particularly like to thank the staff of the following institutes and organisations:

Agricultural Research Council, Weed Research Organization, Oxford.

British Museum (Natural History), London.

Commonwealth Development Corporation, London and Swaziland.

Commonwealth Institute of Entomology, London.

Commonwealth Institute of Helminthology, St. Albans, Herts.

Commonwealth Institute of Mycology, Kew, Surrey.

1

Edward Grey Institute of Field Ornithology, Oxford.

Ministry of Agriculture, Fisheries and Food, Tolworth, Surrey.

MAAF Infestation Control Laboratory, Worplesdon, Surrey.

Plant Protection Ltd., Overseas Technical Service Department, Fernhurst, Surrey.

Tropical Stored Products Centre, (Tropical Products Institute), Ministry of Overseas Development, London.

Woodstock Agriculture Research Centre, Sittingbourne, Kent.

College of Tropical Agriculture, University of Hawaii, Hawaii.

Central Rice Research Institute, Cuttack, India.

Food and Agricultural Organisation, Plant Protection Branch, Rome, and Regional Office for Asia and the Far East, Thailand.

Agricultural Chemicals Inspection Centre, Tokyo, Japan.

National Institute of Agricultural Science, Nishigahara, Japan.

Overseas Technical Co-operation Agency, Tokyo, Japan.

Saitama Prefectual Station, Saitama, Japan.

Faculty of Agriculture, University of Malaya, Kuala Lumpur, Malaya.

Bureau of Plant Industry, Manila, Philippines.

International Rice Research Institute, Los Banos, Philippines.

University of the Philippines, College of Agriculture, Laguna, Philippines..

Rice Production Research Centre, Rice Department, Ministry of Agriculture, Thailand.

Rice Pasture Research and Extension Center, Beaumont, Texas, USA.

The editor would also like to thank CIBA Ltd., Agrochemicals Division, Basle for the production and provision of the colour plates, Mrs H. R. Broad for the many line drawings and all those people who have loaned black and white photographs.

Susan D. Feakin
PANS Manual Editor, 1970

FOREWORD TO SECOND EDITION

PANS Manual No. 3 *Pest Control in Rice* was first published in 1970. During the preparation of this second edition the whole Manual has been revised to give rice farmers an up to date picture of developments in all aspects of pest control in rice. Several changes have been made to make this edition simpler and more practical for those for whom it is intended. In particular literature references to specific methods of pest control have been omitted from the text as it is considered more appropriate that farmers should seek additional advice from local sources rather than from the literature. A list of useful references has, however, been included at the end of each section. A new chapter on machinery has been included.

During the compilation of this edition the *PANS* staff have been assisted by correspondence and literature from a number of organisations and individuals and would like to acknowledge the help of the following:

Agricultural Chemical Experiment Station, Kodaira shi, Tokyo, Japan.

Dr G. L. Campos, Departmento de Arroz, Sueca, Valencia, Spain.

Ciba Geigy (Japan) Ltd., Tokyo, Japan.

Commonwealth Institute of Entomology, London, UK.

Commonwealth Institute of Helminthology, St. Albans, UK.

Commonwealth Mycological Institute, Kew, UK.

Mr P Custodio, Bureau of Plant Industry, Manila, Philippines.

Dr A. K. De Datta, International Rice Research Institute, Manila, Philippines.

Dr V. A. Dyck, International Rice Research Institute, Manila, Philippines.

Du Pont Far East Inc., Japan.

Mr Takeo Endo, Japan Plant Protection Association, Toshima-ku, Tokyo, Japan.

Dr T Furuta, Agricultural Experiment Station, Konosu, Japan.

Mrs P. A. Fynn, Shell Centre, London, UK.

Imperial Chemical Industries Ltd., London, UK.

Dr P. Jones, Centre for Overseas Pest Research, London.

Kumiai Chemical Industry Co. Ltd., Chiyodo-ku, Tokyo, Japan.

Dr K. C. Ling, International Rice Research Institute, Manila, Philippines.

Dr G. A. Matthews, Overseas Spraying Machinery Centre, Imperial College Field Station, Silwood Park, Ascot, UK.

National Institute of Agricultural Sciences, Nishigahara, Kita-ku, Tokyo, Japan.

Nippon Kayaku Co. Ltd., Tokyo, Japan.

ODA Liaison Officers.

Dr S. H. Ou, International Rice Research Institute, Manila, Philippines.

Mr C. Parker, Weed Research Organization, Begbroke Hill, Yarnton, Oxford, UK.

Dr M. D. Pathak, International Rice Research Institute, Manila, Philippines.

Pest Infestation Control Laboratory, Slough, UK.

Dr J. M. Del Rivero, Estacion de Biologia Vegetal, Burgasot, Valencia, Spain.

Rodent Research Center, Los Banos, Philippines.

Sandoz Ltd., Basle, Switzerland.

Shell Kagaku Kabushiki Kaisha, Tokyo, Japan.

Stauffer Chemical Europe S.A., Biggleswade, UK.

Takeda Chemical Industries, Tokyo, Japan.

Mr K. D. Taylor, Pest Infestation Control Laboratory, Ministry of Agriculture
Fisheries and Food, Tolworth, Surbiton. UK.

Texas A. and M. University, Agricultural Research and Extension Center,
Beaumont, Texas, USA.

Dr A. Tinarelli, Ente Nazionale Risi, Milan, Italy.

University of the Philippines, College of Agriculture, Los Banos, Philippines.

Senor J. I. Caballero Ga de Vinuesa, Ministry of Agriculture, Seville, Spain.

Some new line drawings have been prepared by Mrs L. Huddleston and additional
black and white photographs have been provided by various organisations. The
International Rice Research Institute has permitted the use of coloured illus-
trations from their publication *Field Problems of Tropical Rice* by K. E. Mueller
for which we are grateful.

INTRODUCTION

Distribution

Rice is one of the most important cereal foods in the world, in 1974 323,201 thousand metric tons were produced from 136,791 thousand hectares. This cereal is the staple food of millions of people. The estimated area, production and yield by selected regions and countries for 1974 are given in Table 1.

TABLE 1. AREA AND PRODUCTION OF RICE 1974*

Continent and country	Area, 1,000 ha	Production, 1,000 t	kg/ha
Asia			
Bangladesh	9,904	17,222	1,739
Burma	5,059	8,446	1,670
Japan	2,724	15,902	5,838
Pakistan	1,564	3,277	2,094
Philippines	3,467	5,594	1,614
Total Asia	124,186	294,321	2,370
North and Central America			
El Salvador	10	30	2,977
Panama	109	171	1,569
USA	1,040	5,175	4,978
Total North and Central America	1,783	6,871	3,853
South America			
Argentina	83	316	3,821
Brazil	4,378	6,817	1,557
Colombia	362	1,449	4,003
Peru	89	361	4,070
Total South America	5,383	10,158	1,887
Africa			
Egypt	490	2,400	4,898
Liberia	168	208	1,238
Niger	21	43	2,048
Total Africa	4,463	7,595	1,702
Europe			
Greece	20	103	5,029
Italy	190	988	5,209
Spain	61	384	6,295
Yugoslavia	7	31	4,429
Total Europe	398	1,916	4,815
Total World	136,791	323,201	2,363

*Source — FAO (1974). *Production yearbook* Volume 28—1

Within the continents there are striking differences between countries e.g. in Asia, Korea Republic 4,823 kg/ha and Burma 1,670 kg/ha; in Europe, Romania with 2,361 kg/ha and Italy with 5,209 kg/ha and in North and Central America, USA with 4,978 kg/ha compared with Jamaica — 1,050 kg/ha.

It must be remembered that in many parts of Asia rice is planted through necessity in areas not ideally suited to the crop, whereas in most countries where high yields are obtained, the land has been carefully selected because it was considered suitable for rice production. Perhaps the main reason for these wide differences in yield is that the countries of high yield are mainly in the warm temperate zone where the high-yielding and fertilizer-responsive *japonica* varieties can be grown, whereas the yields in tropical countries are, in the main, produced from the hardy but lower-yielding and less fertilizer-responsive *indica* rice varieties. The warm temperate zone has longer days and a more equable climate, also the area's farming techniques are generally more sophisticated. This is not a complete answer since, for instance, in Sierra Leone, farmers from Taiwan have demonstrated that they can, in this tropical region, produce two annual *japonica* rice crops a year from the same field, weighing an aggregate of 1,814 kg. It is evident, therefore, that while the class of rice has a great influence on yield, an equal, if not greater influence is water control and the techniques used in cultivating the crop. It has been proved that the very careful technique for cultivating rice customary in Japan, which results in high yields under tropical conditions, when introduced into India around the year 1952 produced an equally startling result.

Description of the plant

Rice belongs to the tribe Oryzeae and family Gramineae. Most cultivated varieties of rice are in the diploid species *Oryza sativa*, although *Oryza glaberrima* is also widely cultivated in parts of Africa. The latter differs from *Oryza sativa* mainly by a lack of secondary branching of the primary branches of the panicle and on other minor morphological details. Both *Oryza sativa* and *Oryza glaberrima* have 24 chromosomes ($2n=24$).

The rice plant varies in height according to the variety, from a few centimetres to 'floating' varieties exceeding 5 m. The vegetative parts of the plant consist of roots, culms (stems), leaves and panicles. A branch of the plant bearing culm, leaves and roots and often a panicle, is a tiller.

Newly harvested rice usually has a low germination capacity which improves after a period in storage. This seed dormancy may be a function of the plant's photoperiodic sensitivity. The possession of dormancy prevents germination in a sensitive variety of rice until a favourable photoperiod intervenes. On germination, the radicle develops from the base of the grain, quickly followed by two additional roots, all subsequently giving rise to short lateral roots. The main rooting system is composed of adventitious roots produced from the underground nodes. In 'floating' rice varieties, whorls of adventitious roots are also formed from the first three very short nodes, giving the plant additional support. Rice is not an aquatic plant so it has branched roots with hairs.

The culm and leaves develop from the plumule. The culm is more or less erect, cylindrical, smooth and hollow except at the nodes. The number of nodes in the culm ranges from 13−16; usually four internodes elongate. The upper internode (peduncle) is usually the longest and bears the panicle. The tiller produced from the main stem is the 'primary' tiller and is quickly followed by further tillers. The plant habit varies from spreading to compact. The leaves are alternate and borne in two ranks along the stem. Each leaf consists of a sheath enveloping the stem and a flat blade. The uppermost lead or 'flag leaf' of the axis possesses a blade always shorter and broader than that of the lower leaves.

The panicle is usually fairly dense, branched and drooping. The rachella bears the spikelet (Fig. 1) which is laterally compressed, it is one-flowered and articulates below the outer glumes. The two outer glumes are usually shorter than the lemma or flowering glume, whose tip is sometimes prolonged to form an awn. The palea, or upper flowering glume, is similar to the lemma but narrower.

The flower consists of a pistil and 6 stamens composed of 2-celled anthers borne on slender filaments. The grain (caryopsis) which consists of the embryo and the endosperm is tightly enclosed by the lemma and palea (the 'hull'), it is enveloped by the pericarp, which is fibrous and varies in thickness.

Red or purple colouration of many shades is a varietal character and may occur in leaves, nodes, glumes or pericarp.

Classification

The rice plant probably derives from two sources; it is likely that the main origin was in Southeast Asia, and the other in Africa. Cultivation of the crop dates from the earliest age of man and, long before the era of which there is historical evidence rice was a staple food and the first cultivated crop in Asia. The real origin of rice remains a matter for conjecture; it is presumed that the cultivated species have developed from certain wild rices, of which there are still a number.

There are about 25 species of rice, distributed in tropical, sub-tropical and warm temperate regions throughout the world. The most commonly cultivated species is *Oryza sativa*, of which there are, literally, thousands of varieties. The wide differences in characters exhibited by these varieties enables the crop to be grown with success over a wide range of climatic, soil and water conditions; from upland, rain-fed regions to deep undrained swamps; in depths of water (fresh or saline) varying from a few centimetres to 5 m; and in various types of soil.

The cultivated varieties of *Oryza sativa* are divided into groups based mainly on the sterility of the hybrid, since it was observed that crosses of varieties grown in a temperate region and those from tropical zones had a large number of sterile florets. Temperate zone varieties are named *japonica* and tropical ones *indica*. An intermediate basic form is referred to as 'bulu' or *javanica*. The morphological basis for distinguishing these sub-species is somewhat obscure but there are good physiological and genetical grounds for separating the sub-species *japonica* and *indica*, since although the chromosome numbers of both are identical, there is clearly some incompatibility between the genes of the two forms.

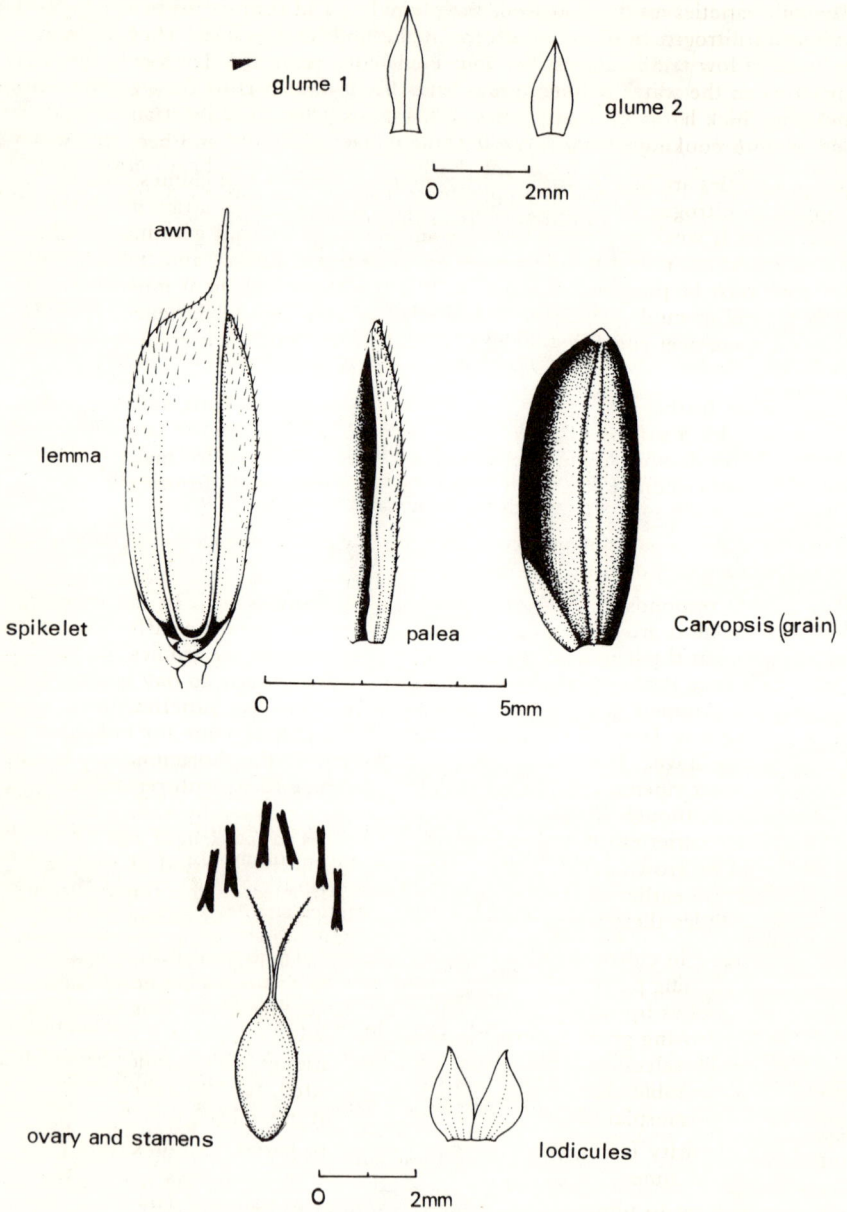

glume 1

glume 2

0   2mm

awn

lemma

spikelet

palea

Caryopsis (grain)

0   5mm

ovary and stamens

lodicules

0   2mm

Fig. 1. A single spikelet of rice dissected to show the various parts. (H.R.B.)

Japonica varieties are short-strawed, high-yielding and respond to heavy applications of nitrogen fertilizer by producing high yields of grain. They are more tolerant of low temperatures than *indica* and may grow and develop faster than *indica* when the water temperature is low. They have narrow dark green leaves and long thick hairs on their glumes. Their grain (Fig. 2) is short and round, and of poor cooking quality since it tends to soften rapidly and become 'mushy'.

Indica varieties are hardy, with slightly pubescent leaves and glumes. They respond to nitrogen fertilizers by an increased production of straw not grain. Their straw is weak and tends to lodge and the crop will not germinate under water, therefore, unless these varieties are grown in a nursery and transplanted, the seed must be pregerminated before it is sown in water, or it must be sown in moist soil or mud. The grain is long, narrow and slightly flattened, it tends to resist some over-cooking, and gives a cooked rice with each grain separate and non-sticky.

Varieties are further divided into two classes; those with a hard, starchy grain and those with a soft, dextrinous grain. The former is the more popular, about 90% of all production being this type; the latter is favoured for making sweetmeats and other confections, and is popular for ceremonial purposes.

Day length

Rice growth responds to changes in day length (photoperiod). In the warm temperate zone it grows in the summer months when there is a difference of up to 4 h between the length of the day and night, while in the tropics the maximum difference is about one hour. Based on their response to day length, rice varieties are grouped as sensitive or non-sensitive. Sensitive varieties flower when the day length is decreasing and when it reaches a critical value for induction of the flowering stage. This inducement of flowering by the shortening day length influences their ripening period, so that they are date-fixed with regard to maturity date, though their growing period can be extended by earlier sowing. Non-sensitive varieties do not respond to differences in day length and consequently can be grown at any season. They require a fixed period of time for maturation and earlier or later sowing has little influence on this time. Among sensitive varieties there are variations in degree of response to photoperiod.

The rice found in cultivated fields usually consists of plants of different types, habits and yielding power. Crop improvement aims at producing new and improved varieties by breeding to combine desirable characters, thus increasing yield and improving grain quality. In short, the main endeavour is directed to the creation by selection and breeding of a few varieties with a wider geographical range. The desirable characters have been enumerated, for rice suited to intensive cultivation by transplantation under tropical conditions. They are:

1. Early maturity (100–120 days from seeding to harvest) to maximise yield per unit of time.
2. Insensitivity to photoperiod, to give flexibility of planting date.
3. Moderate tillering to minimise mutual shading and over-elongation of internodes.

4. Short, sturdy culms to minimise lodging.
5. Small, thick, erect, dark green leaves, to maximise light utilisation.
6. Resistance to the prevailing races of blast disease.
7. Seed dormancy, to avoid germination in the panicle.
8. Moderately firm threshability, to reduce shattering losses.
9. High milling yield.
10. Acceptable shape, size, and cooking quality of grain.

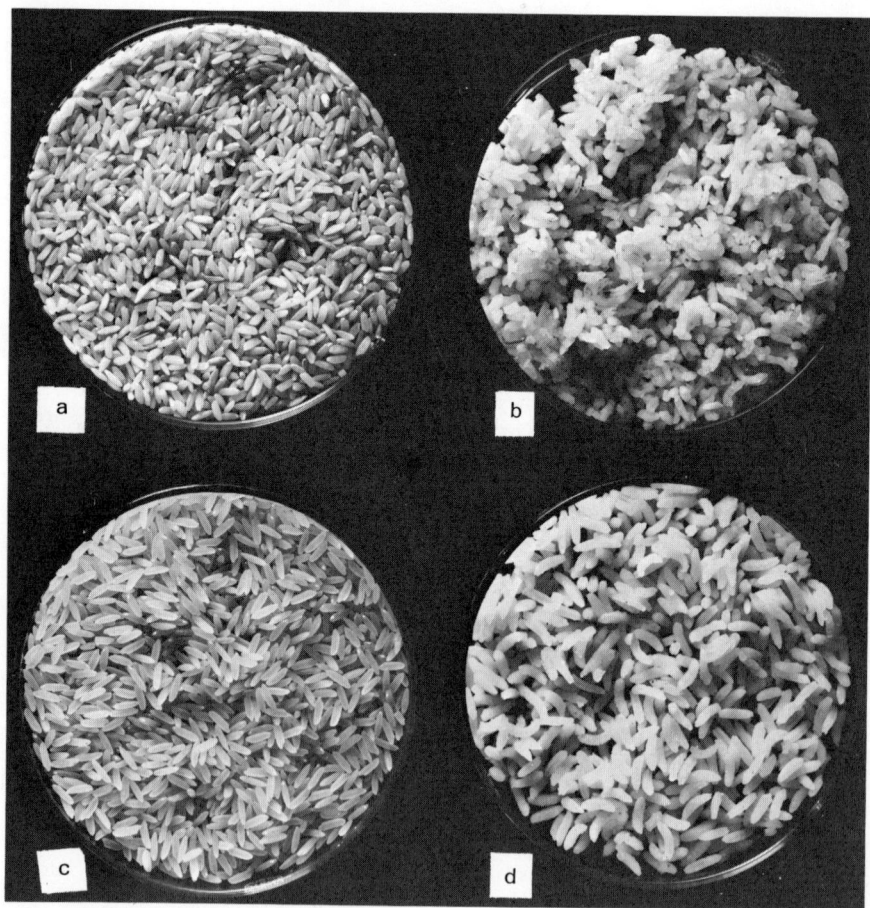

Fig. 2. Short grained rice of the *japonica* type (a) uncooked and (b) cooked showing soft sticky consistency. Long grained rice of the *indica* type (c) uncooked and (d) cooked, showing separate non-sticky grains. (Tropical Products Institute)

It is reasonable to anticipate that the thousands of varieties of rice at present considered necessary to meet the widely differing climatic and soil conditions under which the crop is grown will in the future be replaced by a few varieties capable of producing satisfactory yields under a wide range of conditions.

Although promising results have been obtained by crossing *indica* with *japonica* varieties, there are difficulties with this technique. It has been found that a few *indica* rices are responsive to heavy nitrogen fertilization, also that some *japonica* rice may be cultivated with success under tropical conditions. Promising varieties are found in *japonica* from Taiwan, a number of *japonica* x *indica* in the USA, in dwarf *indica* from Taiwan and the SML varieties from Surinam.

Agronomy

Transplanted

Perhaps the best known of all traditional rice growing systems, and very widely practised throughout Asia. Small rice fields or paddies are often terraced to make maximum use of land area for production. A complex system of bunds helps to control the water in the paddies and lowland varieties of rice are grown in them. Pregerminated seed is planted in nursery beds which are prepared with great care. Before planting seed may be immersed in salt water and only the seed which sinks is used for planting. The selected seed should be dressed to prevent attack from seedborne diseases in the nursery (see Diseases p. 43). In the nursery the seed is broadcast to give a stand of seedlings. Under the Dapog system the seed is scattered onto wet concrete, banana leaves, plastic sheets or matting and covered with banana or other leaves which are removed as the seedlings grow; they are ready to be transplanted in 10—15 days. When ordinary soil beds are used the seedlings are usually transplanted after 30—40 days.

Paddies are ploughed before the rainy season begins or after it has started if buffalo are used, and then puddled to provide a fine tilth after flooding. This puddling is usually carried out by traditional methods, buffalo are driven round and round to trample the soil or used to draw a peg-toothed harrow over and through the mud. In some areas Japanese-developed mechanical puddlers are now being used. In India the land is often levelled again using a bullock-drawn sledge over the mud before the rice is transplanted.

Transplanting is still mainly carried out by hand, but mechanical transplanters are available. Plants for mechanical transplanting are grown in boxes, sometimes under cover, so that the machine is supplied with uniform seedlings.

Transplanted seedlings are removed from the nurseries and should be planted out at regular intervals in rows in the mud, usually more than one seedling per 'hill'. If regular planting intervals are maintained future weed control operations are very much simplified.

Wet sown

The so-called lowland varieties of rice are often sown directly onto mud or into standing water. The seed is usually pregerminated before sowing. This system is practised in both large and small scale fields. In Sri Lanka for example, the fields are cultivated to a depth of 10–20 cm about one month before planting. Secondary tillage consists of puddling the soil after it has been flooded, the wet soil is worked to a depth of about 15 cm and a fine mud is produced. The puddling not only provides a suitable medium for rice to grow in, but also builds a hard pan layer which prevents undue leaking. Pregerminated seed is then broadcast onto the mud surface.

Larger scale wet seeding is practised in the USA and Australia where large fields are levelled, ploughed dry and then irrigated. This helps to control *Echinochloa crusgalli* an important weed of rice fields. The rice is pregerminated and broadcast from the air. Irrigation water is very accurately controlled in these large paddies.

Dry sown

A system which is widely used in southern Africa and Australia, large fields are mechanically graded and cultivated to a fine tilth. Seed is drilled into the dry soil and the irrigation water is let onto the field and carefully controlled throughout the growing season.

Floating rice

In some areas e.g. the central plain of Thailand, it is not yet possible to control the depths of floods caused by the very heavy rains. In these areas floating varieties of rice are grown. They are dry sown and their stems elongate as the water level rises, often to a length of over 4.5 m. The inflorescence is held above the water surface and when the flood subsides the long stems collapse.

Upland

Fields are prepared before the rainy season starts. An assured rainfall over 3 or 4 months is necessary as the crop is entirely rainfed and its water supply is not controlled. The land is prepared by ploughing and harrowing before the rains; weeds and crop debris are ploughed in. When the rains begin the rice seed is broadcast onto the fields, which are enclosed by bunds or levees to retain as much water as possible during the rainy season.

Rotations

In many parts of Asia rice has been planted annually on the same land for centuries. Notably in China and Japan but also in many countries which experience a 'winter' season, one or two crops are reaped in the summer months followed by winter crops such as sweetpotatoes, green vegetables and tobacco. In India where the more or less permanent aquatic conditions of the land preclude cultivation of any crop other than rice the simplest form of rotation is rice followed by fallow.

In some countries which have a winter season, a rotation of grass for a few years after one or more rice crops has proved very satisfactory. Rotation of rice with other crops demands adequate control of irrigation and drainage throughout the year. This degree of control is obtained in many sub-tropical and warm temperate countries, such as Italy, Spain, New South Wales and the USA where rotation of rice with pasture or temporary dry-land crops is successfully operated and the yields are amongst the highest in the world.

Double or multiple cropping

In many countries a substantial increase in rice production is obtained by planting and reaping two or more rice crops a year from the same land. The two essential factors for the multiple cropping are an assured and controlled water supply and a suitable temperature throughout the growing period of the crops. Nearly all the important Asian rice-producing countries have improved or are improving and extending their irrigation systems, so that the area under double cropping with rice is likely to increase considerably in the next few years. This rapid succession of crops is made possible by planting short-maturation varieties (about 120 days from seeding to harvest) or by transplanting at least one of the crops. Under particularly favourable conditions and by observing carefully the most suitable methods of cultivation, phenomenally high yields of rice are possible by multiple cropping of improved varieties. In Malagasy, trials have yielded 17,577—20,088 kg/ha. In India yields of more than 4,484 kg of paddy/ha/year have been obtained without manuring.

Ratooning

Cultivated varities of rice are generally grown as annuals. If the crop is cut and the field is irrigated again, axillary buds develop at the lower nodes from which another crop may be obtained. *Japonica* varieties are usually unsuitable and *Oryza glaberrima* cannot be used for ratoon crops. Experiments in the USA indicate that two-thirds as much nitrogen applied to the ratoon crop will produce about the same grain yield/pound weight of fertilizer as with the first crop. Usually, however, the ratoon crop cannot be expected to yield more than half that of the first crop.

WEEDS

Introduction

Weed competition can be as serious in rice as in any other crop and in the extreme case uncontrolled weed growth can cause complete crop failure. What is not so generally realised is that significant yield reduction may be occurring in spite of the normal weeding practices and in spite of the fact that the crop looks 'normally' healthy. Unlike pests and diseases which cause visible damage, weeds may cause up to 20% crop loss without inducing any obvious symptoms of starvation. Furthermore the conventional hand and hoe weeding practices may not only be inadequate to prevent weed competition (unless carried out very early) but also be directly damaging to the crop.

In transplanted rice, thanks to the weed control effects of water, and the initial advantage that the rice has over the weeds, losses should be minor, but in direct-sown rice losses may frequently be severe and the potential for improving yields by suitable herbicide treatments correspondingly great.

In addition to their direct effects on yield, weeds may also lower the market value of the crop by reducing quality (especially wild and red rices), increase the cost of harvesting, drying and cleaning and also serve as alternate hosts of insects and virus diseases. *Echinochloa* spp. for instance have been associated with the incidence of *Oebalus pugnax* and *Lissorhoptrus oryzophilus*.

Classification of weeds

Grasses

This is the most important and abundant class of weeds in most rice growing systems. In spite of their close relationship to the crop there are now highly selective chemicals for their control but under manual weeding systems there can be difficulty in distinguishing crop and weed in the early stages. Ligule and auricle characters may be useful guides to identification (see Fig. 3). *Echinochloa crusgalli* is almost certainly the most ubiquitous weed of paddy rice (see Fig. 4) but other genera are important, particularly *Leptochloa* spp. (Fig. 5) and *Ischaemum rugosum*. A wider variety of species infests upland rice including *Echinochloa colonum*, *Digitaria* spp., *Eleusine indica*, etc.

Sedges

Many of the sedges are well adapted to the wet conditions of rice cultivation and *Cyperus iria*, *C. difformis* and *Fimbristylis miliacea* are particularly widespread in occurrence (see Figs. 6–8). Sedges and other monocots (unlike the grasses) are generally susceptible to 2,4-D and related growth regulator herbicides and the annual species are not difficult to control. Some important perennial species are mentioned below.

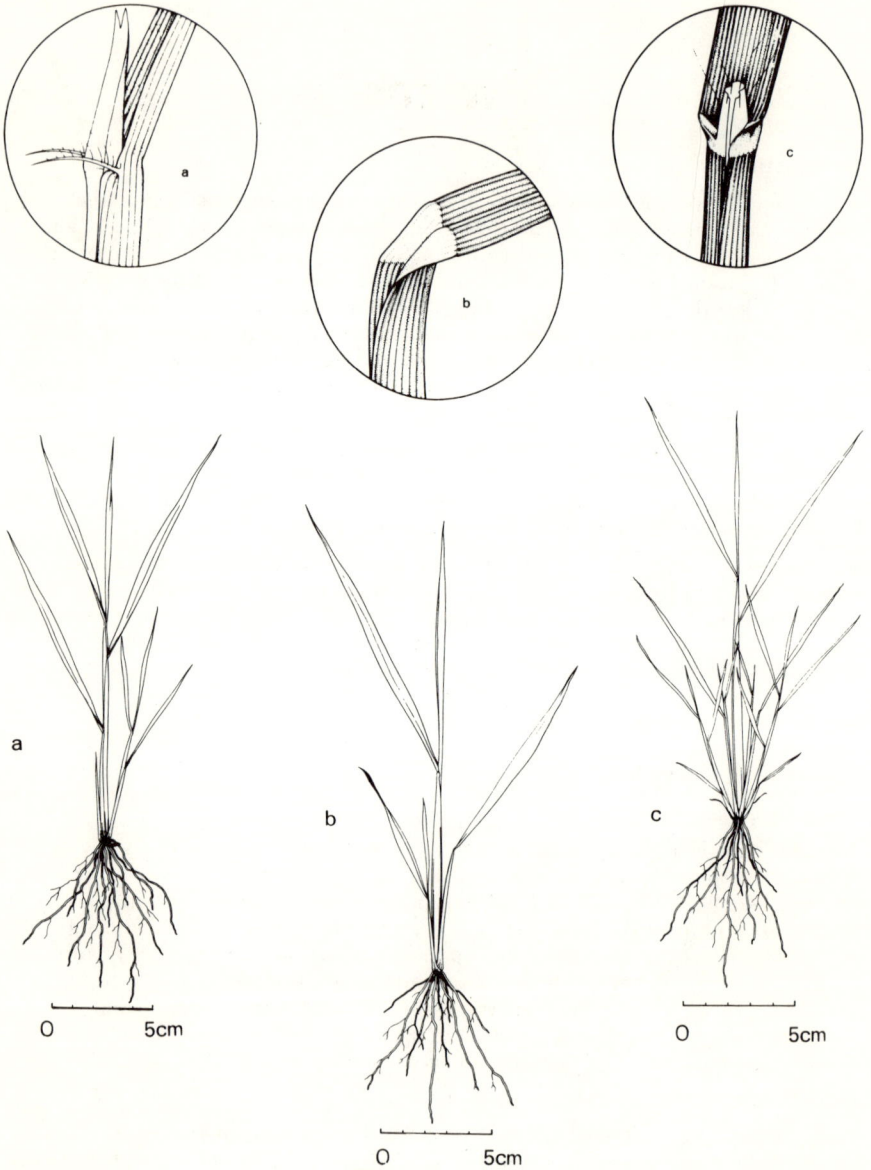

Fig. 3. Young plants of (a) Rice, (b) *Echinochloa crusgalli* and (c) *Leptochloa chinensis* showing their similarities in the seedling stage and (inset, magnified) the ligule areas by which they may be distinguished. (H. R. B.)

Fig. 4. *Echinochloa crusgalli* barnyard grass, upper right the awned type of inflorescence. (H. R. B.)

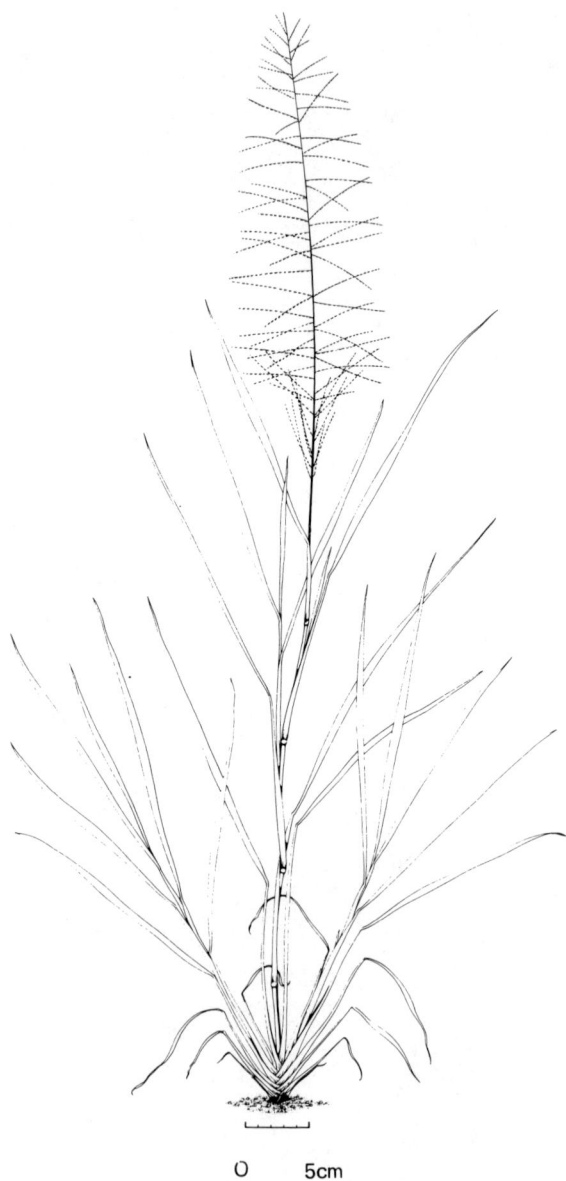

Fig. 5. *Leptochloa chinensis,* A widespread and troublesome weed of flooded rice. (H. R. B.)

Fig. 6. *Cyperus iria* a typical annual sedge with triangular stem (a) Seedling, (b) Mature plant. (H. R. B.)

Fig. 7. *Cyperus difformis* a nutgrass, sedge or umbrella plant with a wide distribution as a weed in rice. (H. R. B.)

O   5cm

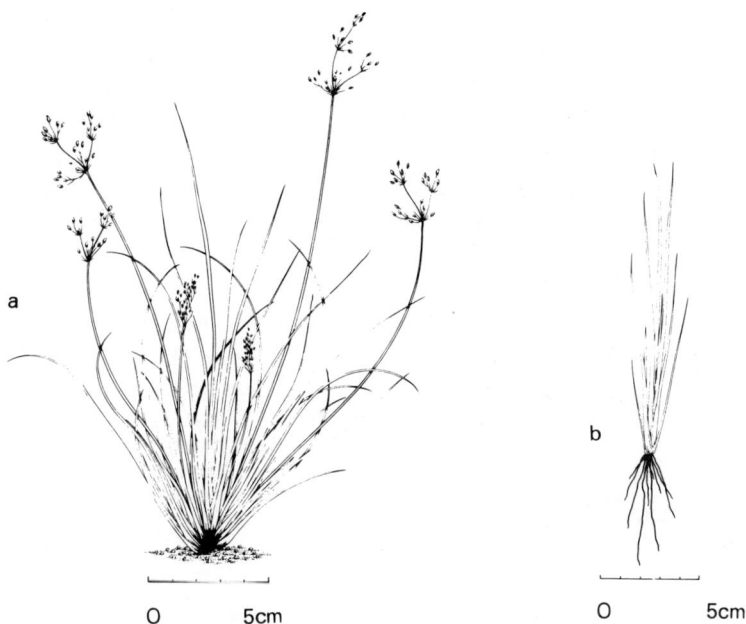

Fig. 8. *Fimbristylis littoralis*, a fine-leaved sedge found in both lowland and upland rice (a) Mature plant, (b) Seedling. (H. R. B.)

Broadleaved species

The range of broadleaved weeds found in wet-sown rice is relatively narrow being confined to more-or-less aquatic species. Examples among the dicotyledons are the *Jussiaea* spp. (Fig. 9), *Ludwigia* spp. (Fig. 10), *Ammania* spp., *Racopa* spp.. and *Sphenoclea zeylanica*.

Among the monocotyledons the most common perhaps is *Monochoria vaginalis* (Fig. 11) with *M. hastata*, *Sagittaria sagittifolia* (Fig. 12), *Heteranthera limosa* and *Eichhornia crassipes* providing more localised problems.

Ferns

Important aquatic ferns are *Salvinia molesta* (previously known wrongly as *S. auriculata*) (Fig. 13), *Azolla* spp. and the creeping *Marsilea* spp. The floating *S. molesta* can form such dense mats as to interfere seriously with land preparation and planting.

Algae

This truly aquatic group of weeds tends to be encouraged by stagnant water conditions. Unicellular green and blue-green algae may not pose any problems but *Chara* spp. and the filamentous genera such as *Spirogyra* can cause trouble.

21

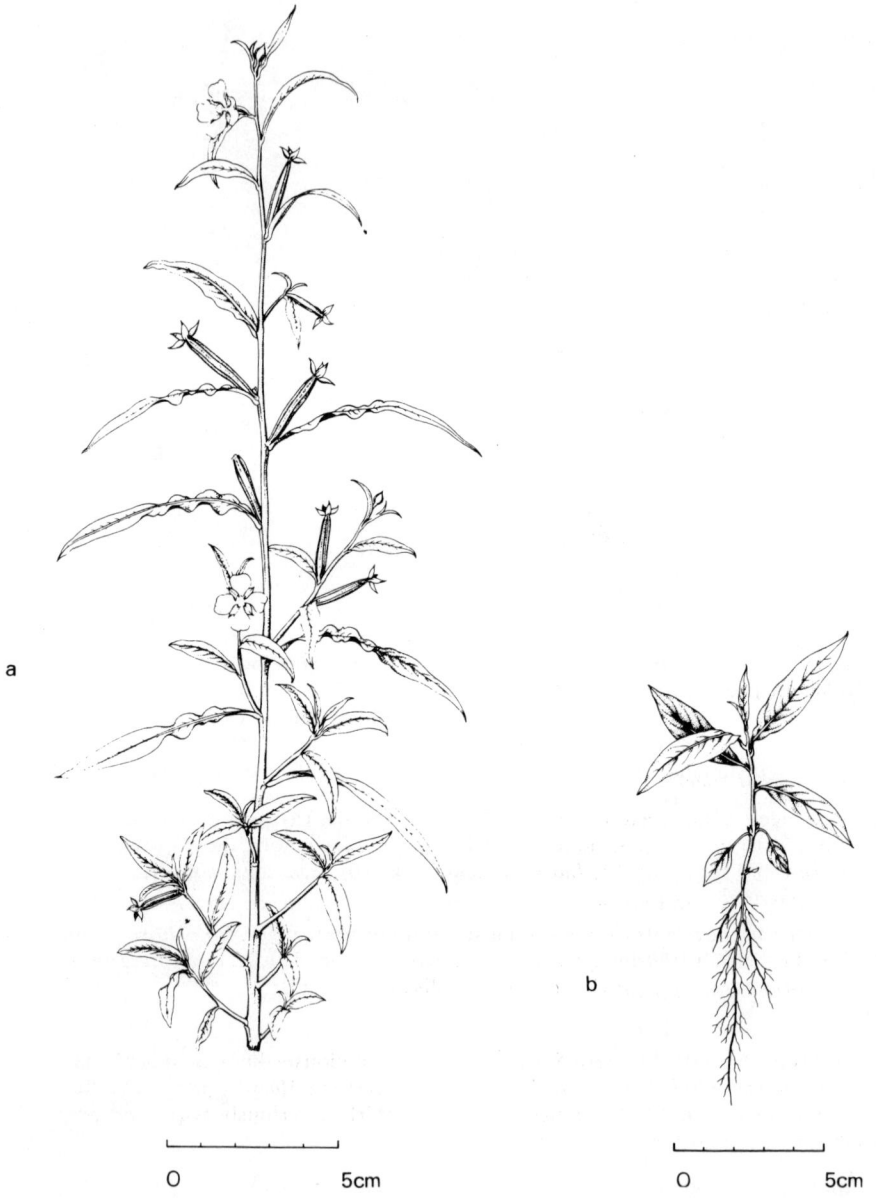

Fig. 9. *Jussiaea suffructicosa*, water primrose, a typical weed of irrigated rice fields (a) Mature plant, (b) Seedling. (H. R. B.)

Fig. 10. *Ludwigia hyssopifolia* seedling.
A broadleaved weed which is fairly easy
to control before rice is transplanted.
(H. R. B.)

0   2cm

Aquatic versus terrestrial

Most of the typical rice weeds, not surprisingly, are aquatic or semi-aquatic in
type. The weeds typical of upland, drained aerobic soils are discouraged by the
wet conditions of lowland rice but may thrive in upland rice or in the early
stages of dry-sown/irrigated crops. It is in these latter crops, therefore that a
much wider range and intensity of weed problems can be expected. In Taiwan
a total of 142 weed species are recorded in rice paddies (25 of these classified as
major). In upland crops 277 species are recorded (26 major). In wet rice the
sedge weeds are usually restricted to the annuals such as *Cyperus difformis* and
C. iria but under upland conditions the much more difficult *C. rotundus* thrives.
The range of grass weeds similarly depends on how aquatic the situation is.
Echinochloa colonum and certain types of *E. crusgalli* cannot germinate under
wet conditions and are restricted to dry-sown or badly irrigated crops. There are,
however, ecotypes of *E. crusgalli* that can germinate under water and hence
flourish even in transplanted rice.

Perennial versus annual

Under traditional rice growing techniques a very large proportion of rice weeds
are annuals. With the use of herbicides, however, the more easily killed annuals
have in certain areas been partially replaced by much more difficult perennial
species and these are becoming of increasing importance. A notable example
among the sedges is *Scirpus maritimus* which is already an extensive problem in
the Philippines and in Italy. (see Fig. 14). It has an underground system of
rhizomes and tubers comparable to that of *Cyperus rotundus* but unlike the
latter, thrives under flooded conditions. A number of other perennials are
particularly troublesome in Japan, including the sedges *Cyperus serotinus,*
Eleocharis acicularis and *Scirpus hotarui* and the Alismataceae; *Sagittaria pygmaea,*

23

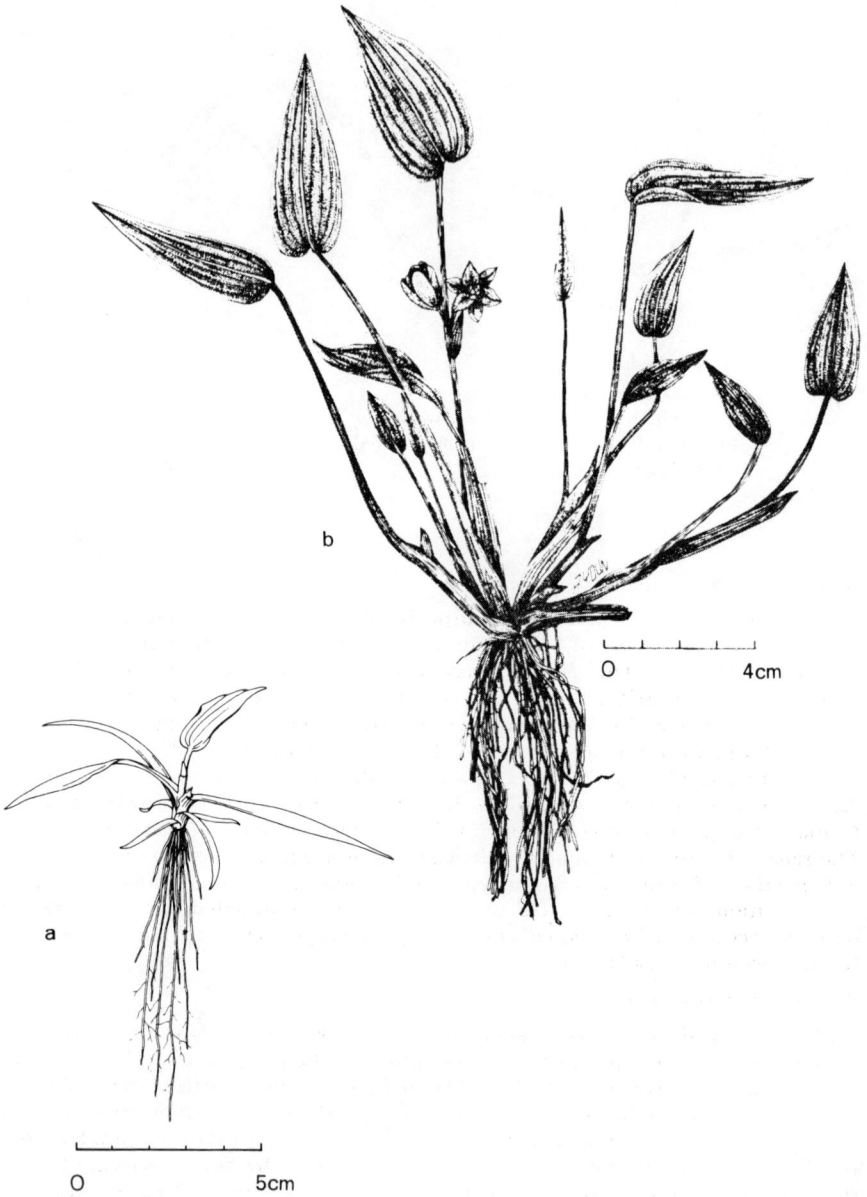

Fig. 11. *Monochoria vaginalis* (a) Seedling (H. R. B.), (b) Mature plant (A. Youn, University of Hawaii). This broadleaved weed can stand deep water.

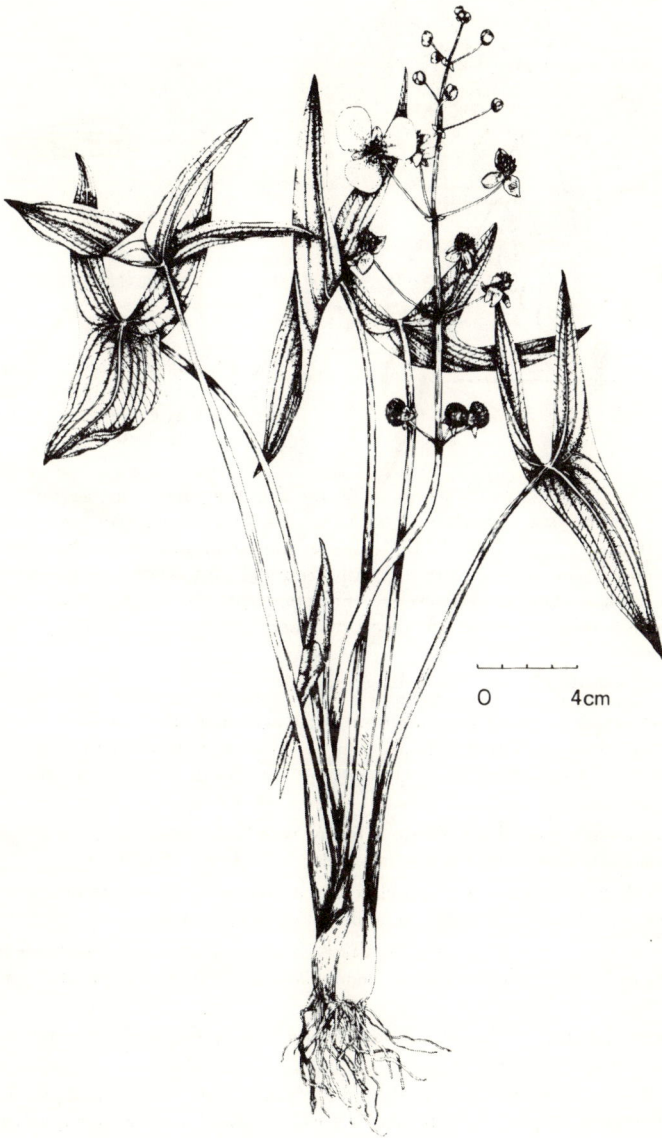

Fig. 12. *Sagittaria sagittifolia* a large broadleaved weed often found in standing water. (A. Youn, University of Hawaii)

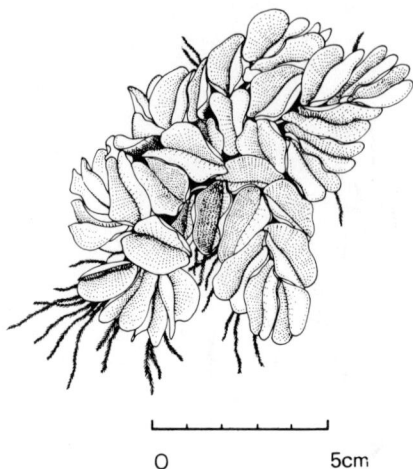

Fig. 13. *Salvinia molesta* a floating weed which is a serious problem in flooded fields and irrigation canals. (H. R. B.)

O   5cm

S. trifolia and *Alisma canaliculatum.* Perennial grasses have not so far increased to the same degree but *Paspalum distichum* and *Oryza longistaminata* are important in places. In all cases the underground rhizome and tuber systems make these weeds more resistant to both chemical and mechanical control and vigilance is always necessary to ensure that such species are not building up, especially when reliance is being placed on chemical weed control.

Prevention of weeds

Although weeds are almost inevitable in most agricultural situations, there are of course procedures which will help to reduce their occurrence and hence the cost and difficulty of controlling them after the crop has been planted. Some of these procedures are as follows:

1. Only clean seed should be planted; many weeds are spread as seeds, particularly grasses and red rice whose seeds are similar to those of rice itself.
2. Keep bunds and irrigation channels clear of weeds as these drop seeds into irrigation water which carries them into the rice crop.
3. Keep tools and machinery used in the rice crop clean. Soil, with weed seeds or pieces of weed which can grow again is easily carried from field to field on machinery.
4. Keep livestock out of the rice fields as much as possible as they will also carry weed seeds into the crop area.
5. Be sure that before the crop is sown or transplanted the ground is clear of weeds; this will help in establishing a good stand of rice, as competition between crop and weeds is critical in the early stages. Careful tillage with complete inversion of the furrow slice will go a long way to removing many weeds which have grown since the last harvest. The use of a herbicide at this stage should be considered.

26

Fig. 14. *Scirpus maritimus* a sedge which is an important perennial weed of rice.

6. Make fertilizer applications at the time they will do most to help the crop and not when they will increase the weed competition e.g. phosphate or nitrogen applied directly to rice stimulates the growth of many weeds; grasses, broad-leaved and algae. Phosphate applied before dry seeding stimulates the growth of young grass plants. Where fields have a history of severe infestations of *E. crusgalli* phosphate should be applied to a crop in the rotation other than rice, or just before early flooding of rice. When herbicides are used to control *Echinochloa* spp. phosphate can be applied to young rice in the early stages of growth so that the rice will obtain maximum benefit. Early to mid-season applications of nitrogen are beneficial to the rice crop, but may increase weed competition to the extent of decreasing the crop yield. If nitrogen is applied to the rice crop, its effect on the rice will be maximised if weeds have previously been removed from the crop.

7. Rotation into alternative crops will help to prevent the build-up of the typical semi-aquatic rice weed flora.

Weed control in transplanted rice

Pre-planting operations

Mechanical land preparation in the paddies should destroy as many weeds as possible before transplanting. Up to 7 cultivations may be needed using a bullock-drawn wooden plough but tractor-drawn ploughs and the small motorised rotovators widely used in Japan are more efficient. In the preliminary ploughing the weeds left from the previous crop are turned in and buried. During puddling weeds which have grown since the ploughing and have not been burned or up-rooted are uprooted and buried in the lower layers of mud. These weeds decompose by anaerobic action to form ammonium compounds which are retained in the soil and used by the crop.

The time taken for land preparation by mechanical methods before direct seeding or transplanting the crop is often determined by the weeds and amount of crop debris in the field; this is particularly true in areas where rice is grown in rotation with other crops. In some areas preparation may entail slashing down the weeds and flooding the fields for several weeks to rot them before cultivations can continue. As multiple rice cropping increases with the development of new varieties and better water control, the time factor, especially with regard to planting date becomes more important. A technique involving a non-selective contact herbicide has now been developed and is being used in Sri Lanka, Malaysia, Japan and the Philippines. The herbicides involved include paraquat and dalapon. The latter is used 4 days before paraquat where perennial grass problems are particularly serious. Paraquat was originally used to kill *Salvinia molesta* in irrigation canals in Sri Lanka. It was also found to kill other weeds and make subsequent cultivations easier.

An alternative to dalapon now available for control of perennial weeds is glyphosate. This has the same valuable characteristic as paraquat, of being inactive in soil and is therefore completely safe for the crop, but it may prove too expensive for general use.

Post-planting weed control

Water

The correct use of water should ensure that all the normal terrestrial weed problems are totally prevented in transplanted rice. The anaerobic conditions prevailing in soil under 5—10 cm water inhibit virtually all but the specially adapted aquatic species. Any reduction in water level, exposing the soil surface, if only for a short time, will lead to more aerobic conditions and hence germination of weed seeds. Similarly unevenness of the paddies resulting in areas inadequately flooded will lead to patches of problem weeds. Conversely in low spots, the depth of water may be too great for the health of the rice. Rice does not need or even benefit from flooding. The one essential contribution of the standing water is weed control. If it can be well controlled, relatively little may be needed to achieve the weed control function.

Mechanical

Where weeas develop in transplanted rice, because of inadequate water control or because of the presence of aquatic plant types, control may be possible on the smallest scale by hand pulling or by treading into the mud. For the somewhat larger scale or more serious infestations there are many small hand and machine powered weeders such as the Japanese rotary weeder which are run or pushed along between the rows. The critical time for weed competition varies with the variety of rice grown. Most varieties should be weeded as often as necessary within the period 10—40 days after transplanting. If hand weeding and rotary weeding are used, the rotary weeding should be done first. If labour shortage means only one weeding is possible this should be done early, not later than 30 days from transplanting.

Chemical

Where mechanical methods cannot be successfully applied there are now a wide range of selective herbicides available that may be applied within a few days of transplanting and which provide residual pre-emergence control of weeds and so completely prevent weed competition. The lack of treading and soil disturbance may also be beneficial. The herbicides available now include compounds that are not only highly effective and selective but are also of low mammalian toxicity, relatively inexpensive and easily applied. Granular formulations can be distributed by hand and no special application equipment is needed.

The basic compounds to be considered are 2,4-D or MCPA esters which when applied as granules soon after transplanting provide selective pre-emergence control of annual grasses as well as many annual sedge and broadleaved species. Other compounds which are used in combination with or instead of 2,4-D to provide more prolonged or reliable control include EPTC, nitrofen, benthiocarb, butachlor, molinate, trifluralin, TCE-styrene, etc. In Japan these various compounds and mixtures have now replaced PCP which had disadvantages of short residual effect and toxicity to fish. The relative properties of these various compounds are summarised in Table 2.

Under good cultural conditions the use of these compounds may not be justified as a routine measure before weeds are seen. Where weeds have to be treated

TABLE 2. HERBICIDES AND THEIR USE IN RICE

Herbicide	Rate (kg a.i./ha)	Type of crop*	Comments
For land preparation			
Paraquat	0.5	T W }	Used to destroy all above-ground vegetation. After 3 days burn or cultivate and flood. Transplant or sow 2–5 days later.
	0.5	D	May be used to destroy emerged weeds in stale seedbed immediately before sowing.
Dalapon	3	T W }	Used 3 days before paraquat (above) where necessary to improve control of perennial grasses.
	8–15	T W D }	For more complete control of perennial grasses but 4 weeks or more delay may be necessary before sowing, for soil residues to decline.
Glyphosate	1–4	T W D }	For excellent control of most perennial weeds. Must be applied to well established vegetation. Allow 4 days for translocation before cultivation and/or paraquat. No delay necessary before sowing/planting.
For post-planting weed control (In approximate chronological order of development)			
2,4-D } MCPA }	0.5–2	T	As granules 2–5 DAT† for most annual weeds, including grasses.
2,4-D MCPA 2,4,5-T Fenoprop }	0.5–1.5	T W D }	As sprays after tillering but before shooting for most annual dicot and sedge weeds. MCPA and 2,4,5-T safer than 2,4-D. Fenoprop for *Scirpus maritimus*. Risk of damage greater within 7–21 days after nitrogen application. Drain before spraying. Reflood within a few days.
Propanil	3–6	T W D }	Post-em. for young weeds, especially grasses. Expose by draining if necessary, and re-flood within a few days. May scorch rice over 30°C. DO NOT SPRAY WITHIN 10 DAYS OF CARBAMATE OR ORGANOPHOS-PHORUS INSECTICIDES. Can be mixed with 2,4-D, etc but some other herbicides interact.
Nitrofen Fluorodifen Chlornitrofen }	2.5–4	T	As granules 2-5 DAT for annual grasses, and some other species. May be combined with propanil for later application.
Nitrofen Fluorodifen Chlornitrofen }	2.5–4	D	Spray to seedbed 0–3 DAS† for residual pre-em control of annual grasses.

TABLE 2—continued

Herbicide	Rate (kg a.i./ha)	Type of crop*	Comments
Nitrofen Fluorodifen Chlornitrofen	2.5—4	D W	Spray early post-em. for control of annual grasses. Some scorch of the crop may occur.
Molinate	3—5	T W (D)	Spray on to seedbed and incorporate immediately to 5—10 cm for annual grasses only. (May not be safe on dry sown.)
Trifluralin	0.6—0.8	T	As granules 2—5 DAT for annual grasses *only*. Usually in mixture with MCPA or 2,4-D.
EPTC	1.75	T	As trifluralin.
Butachlor	1.6	T	As granules 2—5 DAT for annual grasses and sedges.
	1.6	W	6—8 DAS for annual grasses and sedges. Soil saturated but not flooded for 3 days after application.
	1.6	D	As pre-em. spray (0—3 DAS) for annual grasses and sedges.
	1.6	D W	Early post-em. for residual control of annual grasses and sedges ($\pm$ propanil).
Benthiocarb	1.5—4	T W D	Sprayed on to seedbed and incorporated 5—10 cm deep for annual grasses and some dicots.
	1.5—4	T W	As granules 2—5 DAT or DAS for annual grasses.
	1.5—4	W D	Early post-em. for residual control of annual grasses ($\pm$ propanil).
			Note benthiocarb/propanil mixtures may show synergistic action. Hence greater chance of damage especially on certain rice varieties including IR8.
TCE-styrene	0.5—0.75	T	As granules 3—5 DAT for annual grasses *only* (usually in mixture with 2,4-D).
Piperophos + dimethametryn	0.75—1.25	T	As granules 2—5 DAT for annual grasses, sedges and dicots.
		W D	Spray or granules early post-em. for annual weeds ($\pm$ 2,4-D) but some risk of crop damage.
Dinitramine Butralin	0.4—0.8 2	T	As spray or granules 2—5 DAT for annual grasses *only* (cf trifluralin).
Butralin	2	D	As pre-em. spray 2—3 DAS for annual grasses *only* (cf trifluralin).

TABLE 2—continued

Herbicide	Rate (kg a.i./ha)	Type of crop*	Comments
Oxadiazon	0.6	T	As granules 2—5 DAT for annual grasses and some dicots.
	0.5—1.5	D	Pre or early post-em. for annual grasses and some dicots. Some crop scorch may occur.
Bentazon	1—2	T W D	Post-em. for dicots and sedges including *Scirpus maritimus* and *Cyperus esculentus* but not *C. rotundus*.
Fluoronitrofen Chlormethoxynil	2.5—4	T W D	As nitrofen etc.
Herbicides still under development			
Kue 2079A	1.2	T	2—5 DAT for residual pre-em. or early post-em. control of broadleaved weeds.
		W D	For early post-em. control of broadleaved weeds after rice has 1.5 leaves.
K 223 K 1441	7—10 7—10	D	Incorporate to 10 cm pre-planting for control of *Cyperus rotundus* and annual grasses.
Cyperquat	3—4	T W D	Post-em. for control of sedges including *C. rotundus* and *C. esculentus*. High mammalian toxicity.

*T = Transplanted; W = Wet-sown; D = Dry-sown
†DAT = days after transplanting; DAS = days after sowing

after emergence, most of the above compounds can still be used, again as granules or sprays applied into the water, but 2,4-D, MCPA, etc will no longer kill grasses, and control of water level may be important for the best results with the others. Propanil may be resorted to if necessary at a later stage but the water level must then be lowered and there are various other disadvantages with propanil (see under wet-sown) which mean that it should be regarded as an emergency rather than a routine treatment.

For broadleaved and annual sedge weeds which are not controlled by residual treatments, 2,4-D, MCPA, etc may be safely used as sprays on to the exposed weeds (drain if necessary) provided the rice has fully tillered but not yet begun to shoot.

In all cases some final hand-weeding should be done to remove surviving weeds and prevent seeding.

Special problems

Perennial sedges have built up in areas where herbicides have been used for some years. They are tolerant to most of the herbicides listed above but can fortunately be controlled by fenoprop or by bentazon as post-emergence treatments.

Algae may also thrive in transplanted rice crops. In the Philippines it has been noted that carbaryl insecticide provides some selective control at early stages (and in the nursery). In India the problem may be severe in poorly-drained transplanted crops. Copper sulphate, copper oxychloride or cuprous oxide applied as sprays, or added to the paddy water are very effective, but the dose must be kept small owing to potential crop damage. Control measures are not often taken because of the very high cost of copper. Sprays of 0.67—1 ppm fentin acetate have also been found effective and non-toxic to fish as has granular nitrofen at 1 kg/ha.

Control of all algae may not necessarily be desirable as some blue-green types are believed to fix nitrogen and have some beneficial effect on the crop.

Aquatic species are well adapted to grow in transplanted rice and apart from the sedges, algae and rooted dicots already referred to, there may be problems from the floating species including *Pistia stratiotes, Eichhornia crassipes, Lemna polyrhiza* (Fig. 15) and the ferns *Salvinia molesta* and *Azolla* spp. There is conflicting evidence on the damaging effect of *Azolla* but the others are generally thought to be detrimental and have to be removed beforehand so far as possible as control other than by hand-weeding is difficult to achieve once the crop is planted. Paraquat has been particularly successfully used for pre-planting control of *S. molesta*.

Nurseries

Nurseries for transplanted rice are most often wet-sown and reference may be made to the next section. The areas are generally small and hand-weeding is often all that is required but in order to reduce the risk of weeds being transplanted in mistake for rice propanil may be used especially to control *Echinochloa crusgalli*. For the purpose of distinguishing 'red rice' (see section on dry-sown

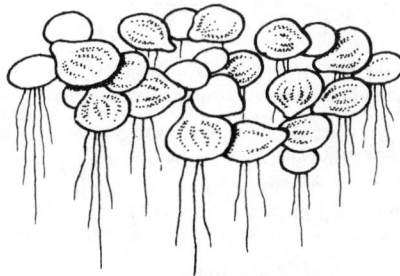

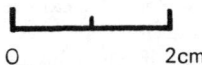

|_____|_____|
O   2cm

Fig. 15. *Lemma polyrhiza* a small floating weed which forms a bright green scum on the water in rice fields. (H. R. B.)

rice for more detail) the crop variety has sometimes had purple foliage bred into it, so that all green rice can be assumed 'red' and discarded before transplanting.

Weed control in wet-sown rice

Pre-planting operations

These are directly comparable to those for transplanted rice but require even greater care in the final levelling of the land as the initial depth of water will be even more critical for the well-being of the germinating rice seedlings.

Planting

In order that the rice may grow away under saturated or submerged conditions, the seed is usually pre-germinated for 24—48 h under aerobic conditions. It can then continue its development even in shallow water. In small scale farming the seed is normally scattered on to saturated mud with little or no standing water. The water level is then gradually raised as the rice grows. In more sophisticated, large scale operations, particularly in cooler, temperate regions, the seed is scattered from the air into shallow standing water 5—10 cm deep.

Post-planting weed control

Mechanical

As the seed is rarely placed in rows there are much greater difficulties to hand or mechanical weeding, but the row weeders are sometimes used to destroy strips (weeds and rice) and so create rows. The advantage of so destroying more than half the sown rice is not altogether clear.

Weeds are not usually so completely controlled by the water in direct wet-sown rice because of the initial shallowness of flooding and the greater likelihood of exposed, aerobic soil conditions. Furthermore the rice is very little ahead of the weeds and competition is liable to be more serious.

Chemical

The fact that the rice and weeds are germinating at the same time considerably restricts the range of herbicides that can safely be used.

Those that are safe at the time of rice germination include benthiocarb and molinate. Either of these compounds can be used as soil-incorporated pre-sowing treatments. Alternatively they can be applied to the water a few days after sowing. At this stage other herbicides that can be safely used include butachlor and substituted ethers such as nitrofen, fluorodifen and chlornitrofen.

If no application is made before the weeds reach the 3—4 leaf stage it may be necessary to use propanil to control annual grass weeds. Before propanil is applied, the water level must be lowered to expose the weeds. It must then be raised again to prevent new germination of weeds. Unfortunately propanil cannot be used within 10—14 days of many insecticides (especially carbamates and organophosphorus compounds). It is also liable to be affected by climatic conditions, sometimes causing scorch at temperatures over $30°C$ and may be washed off by heavy rains. *Leptochloa* spp. are rather more tolerant than

Echinochloa spp. and application has to be earlier or at a higher dose for the best results.

For control of broadleaved weeds and annual sedges, a single application of 2,4-D, MCPA, 2,4,5-T or fenoprop a few weeks after sowing may be all that is required, but care is needed in the selection and timing of the treatment to reduce the risk of crop damage and local recommendations should be closely followed. Tropical *indica* rices are somewhat more tolerant than *japonica* and might be safely treated earlier, but generally the rice should be well tillered and not yet shooting at the time of treatment.

For perennial sedges, fenoprop may be adequate but bentazon has shown promise on a wider range of species.

Weed control in dry-sown/irrigated rice

Pre-planting operations

No specialised procedures are available for control of weeds in this class of crop, but pre-planting cultivations must be designed to destroy so far as possible all old vegetation especially perennial species. If time, soil and climate allow, deep dry season cultivations may be particularly valuable for the control of perennial weeds such as *Cyperus rotundus* and *Oryza longistaminata*. Again if the climate allows it may be possible to prepare a seedbed, wait for rain to cause weed seed germination and then re-cultivate shallowly or spray paraquat before sowing into a relatively weed-free seed bed.

Pre-planting chemical treatments may also be valuable for control of the perennial weeds, and glyphosate is an outstanding new compound for this purpose. It is active on virtually all the major perennial weeds including *C. rotundus* and *O. longistaminata* but is unfortunately an expensive compound that will not be economic in all situations. Dalapon remains a cheaper alternative but unlike glyphosate and paraquat which leave no active soil residues, dalapon has a moderate period of persistence which must be taken into consideration, especially under dry soil conditions.

Post-planting weed control

Water

Following dry-sowing, the crop may be irrigated but it must be immediately drained again to allow germination of the rice. The aerobic conditions so created also allow germination of all normal terrestrial weeds as well as aquatic types and hence the weed problems in dry-sown rice are generally a great deal worse than those in wet sown crops. Irrigation may after a few weeks take the form of continuous submergence which will drown some thoroughly terrestrial species but a great many will be able to continue growing under flooded conditions once they are established.

Mechanical

Even if the crop is sown in rows, these are not usually wide enough for inter-row mechanical weeding and non-chemical weeding in this class of rice usually involves hand or hoe weeding. Unfortunately the amount of treading and soil disturbance involved in intensive hand-weeding can cause almost as much damage as the weeds themselves.

Herbicides are therefore particularly important for this type of crop.

Chemical

The possibilities for chemical control in dry-sown rice are more or less the same as those for wet-sown, but pre-planting molinate is not so well tolerated by rice that is drilled into the treated soil. The safety of molinate can be increased by treating the crop with a seed dressing of herbicide antidote (see under wild and red rice) but the cost and inconvenience of the treatment make it impracticable for general purposes.

Drilling seed into the soil makes it somewhat more tolerant of herbicides applied as pre-emergence treatments. Consequently butachlor, and the substituted ethers (nitrofen, etc) can be applied immediately after sowing without waiting for rice germination. Oxadiazon is also better tolerated. Of all these compounds, butachlor is perhaps the most widely used. This provides control of annual grasses and sedges for several weeks but post-emergence treatments are usually needed for control of broadleaved species and for grasses too if flooding is not introduced before the residual effect has worn off. MCPA and 2,4-D are therefore commonly applied after tillering and propanil is also widely used post-emergence either in addition to or instead of a pre-emergence treatment.

As propanil has no residual effect in the soil it must be followed within a few days by permanent flood or else a residual herbicide such as butachlor or benthiocarb mixed with it. Alternatively a second propanil application may be made shortly before flooding.

Certain limitations and risks of propanil and of 2,4-D and MCPA are mentioned under wet-sown rice (q.v.).

Special problem — wild and red rices

The plant known as 'wild rice' in N. America (*Zizania aquatica*) is not a weed of the rice crop, but there are a number of wild *Oryza* species which do create problems — particularly in dry-sown/irrigated rice. This system allows optimum germination and then provides the wet or aquatic conditions under which they flourish best. The most important species in Africa are probably *O. barthii* (an annual, previously known as *O. breviligulata*) and the rhizomatous perennial *O. longistaminata* (previously known as *O. barthii* or *O. perennis*), both of which are widespread in West Africa but do not occur in other continents. *O. glaberrima* the traditional cultivated species of West Africa may also be regarded as a weed where it is being replaced by *O. sativa*. *O. punctata* is a small-seeded African species particularly troublesome in Swaziland. In Asia and S. America, the main problems are from non-rhizomatous forms of *O. perennis*, and from *O. rufipogon*. The latter are closest of all to *O. sativa* and many taxonomists regard them as subspecies or varieties (e.g. *O. sativa* ssp. *fatua* or var. *spontanea*). All these forms closely related to the crop are generally referred to as 'red rice', a term which can also include types which are definitely *O. sativa* but happen to have a red pericarp. Apart from this red pericarp and usually having awns too, virtually all the wild and red rices have two further important characters in common — seeds which shatter before harvest, and prolonged dormancy. Hence even if the grain is edible (as it is in the case of most 'red' rices), the yield is reduced due to shattering and the problem is perpetuated by seeds carrying over to the next crop. Although the 'red' grains are edible they are undesirable for many markets and the price of the produce may be correspondingly lowered.

In wet-sown crops the thorough saturation of the soil before planting, combined ideally with a smearing of the wet soil surface to make it even more impermeable to oxygen, helps to prevent germination.

In dry-sown crops, there is little that can be recommended for most of the red rices other than ensuring the purest possible seed, combined with intensive hand roguing as the plants come into ear.

Selective control by chemicals can only be achieved with the aid of herbicide antidotes. One such compound, 1,8-naphthalic anhydride used as a seed dressing at 0.5—1% by weight has been shown to protect rice against several herbicides including molinate, alachlor, benthiocarb and perfluidone, so that higher than normal doses can be used and some selective control obtained. This technique has been particularly successful against *O. punctata* in Swaziland using the herbicide alachlor.

Weed control in floating rice

Floating rice is usually established as a dry-sown crop in river valleys where the flooding is uncontrolled and may eventually reach several metres depth. Early weed control may be achieved as for normal dry-sown/irrigated rice but eventually the depth of water should suppress all but a few aquatic species. A particular problem in this class of rice is the rhizomatous perennial wild rice *O. longistaminata* which survives under the moist soil conditions of the fallow period. Dalapon has been used to some extent but without great success. Glyphosate has shown promise but it is not yet clear what dose will be required for reliable control.

Weed control in upland rice

Pre-planting operations

As for dry-sown/irrigated (q.v.)

Post-planting weed control

Mechanical

Upland rice is generally the least productive type of crop, yet with adequate rainfall, very high yields can be obtained. The low yields are very often due to the severe difficulty of controlling weeds adequately. Most upland rice farmers rely on hand-weeding with some hoe-weeding and occasionally the rows are spaced widely enough for an animal drawn inter-row weeding.

Traditional varieties have generally been tall and leafy providing effective shade after some weeks. Newer shorter varieties do not always provide such effective shade and weeds can outgrow the crop and create greater problems than ever. Hence the urgent need for cheaper and more selective herbicides.

Chemical

Herbicides for upland rice are basically the same as those for dry-sown/irrigated, but there may be the need for extra treatments to provide more prolonged control, in the absence of the drowning effect of flooding.

Special problems

Cyperus rotundus is discouraged by, repeated flooding but thrives under the seasonally wet conditions suitable for upland rice. Control of this and most other perennial weeds has to be concentrated in the fallow period. Deep dry-season cultivations and/or the use of herbicides such as glyphosate or 2,4-D may be useful. More recently, however, three experimental herbicides have shown promise for selective control of *C. rotundus* in rice. These are K 223 (N-(α,α-dimethylbenzyl)N′-p-tolylurea) and K 1441 (N-(α,α-dimethylbenzyl)N′-methyl-N′-phenylurea) as pre-planting or pre-emergence treatments and cyperquat as a post-emergence treatment.

Striga spp., especially *S. asiatica* and *S. hermonthica* are root parasites attacking upland rice in parts of West Africa, India and Southeast Asia. The emerged shoots can be killed by 2,4-D, MCPA, etc but the weed is late to emerge and by that time much of the damage has already been done and it may no longer be safe to use a growth-regulator type herbicide. *Striga* in sorghum and maize can be usefully suppressed by application of nitrogen fertilizer and it is probable that the same is true in rice. There is little information on the relative susceptibility of different varieties of rice but, again by analogy with sorghum, there may be some usefully resistant varieties.

Application of herbicides

Unlike insecticides and fungicides which may still be safe on crops when used at many times their recommended doses, herbicides are all intrinsically phytotoxic and must be applied as carefully and accurately as possible. Even a doubling of dose as occurs with overlapping spray swathes may cause significant damage. It is also important that local recommendations are followed as the safety of the herbicide may be affected by local conditions of soil, climate or variety.

Herbicides are formulated as sprays or as granules. Spraying involves a diversity of equipment from aeroplanes to tractors to knapsack sprayers. Any of these may be suitable for dry-sown and upland crops or for pre-planting treatments but in flooded conditions tractors cannot be used and knapsack sprayers may be heavy to handle. Under these conditions, granular formulations are particularly suitable. On a small scale no application equipment at all is required for granules, any imperfections in hand-distribution are made good by diffusion and re-distribution in the water. As there is no dilution step, there is also less danger of miscalculation of dosage by the less educated farmer. Granular herbicides may be somewhat more expensive than liquid formulations but the above advantages will often outweigh the extra cost and local manufacture or formulation should be encouraged to help reduce costs.

In addition to their possible phytoxicity to the crop, herbicides may also be dangerous because of their mammalian toxicity. The manufacturers' precautions indicated for handling the chemicals safely should be followed and care taken to dispose of any contaminated vessels or left-over sprays where they cannot harm other crops, fish or livestock.

Biological control

A fungal anthracnose disease, *Colletotrichum gloeosporioides* (Perz.) is now at an advanced stage of development for possible commercial use against *Aeschynomene virginica* in USA.

In Japan it is hoped that shrimps of the genus *Triops* may be useful as a means of weed control in transplanted rice. They burrow in the soil surface disturbing young seedling weeds. They can, however, also be damaging to direct-sown rice.

Preventing the build-up of tolerant weeds

No one herbicide nor even a combination of several herbicides will control all weeds and there will almost invariably be survivors of chemical weed control. With repeated use of the same chemical(s) there is an inevitable increase in these 'tolerant' weeds which may have previously been unimportant and even unnoticed. The perennial weeds are particularly important in this respect e.g. *Scirpus maritimus, Cyperus rotundus, Paspalum distichum*, etc. The wild and red rices are other examples which are likely to increase especially where transplanted rice is replaced by direct (dry) sowing and weed control by water is replaced by chemical control. 'Rotating' herbicides i.e. using different compounds or combinations in successive crops on the same land is a partial answer, preventing particular annual weed species from building up, but the only certain way of avoiding such problems is always to retain an element of hand-weeding or hand-roguing whenever it is feasible, and to use other hygienic and cultural systems which will reduce weed populations both during and between crops.

Bibliography

CHANG, W. L. and DE DATTA, S. K. (1974). Chemical weed control in direct-seeded flooded rice in Taiwan. *PANS* 20 (4): 425–428.

CHIAPPARINI, L., BALDACCI, E., MOGLIA, C., GOLDBERG, F. L. and VANDONI, M. V. (1964). Sull'impiego del trifenil-acetato di stagno come alghicida, I–II. (On the use of tin triphenyl acetate as an algicide, I–II). *Il Riso* 13 (3): 227–238.

CHISAKA, H. (1975). Trends in weed control for lowland rice in Japan. *Japan Pesticide Information* 22: 11–15.

CONSTANTINESCO, I., EL TONBARY, A. A., FARDUKY, S. T. and LODGE, R. W. (1964). Proposals for land improvement of the Soan Valley. *FAO Expanded Progress Technical Assistance Report*, 1963. pp. 25.

DANIEL, J. T., TEMPLETON, G. E., SMITH, R. J. and FOX, W. T. (1973). Biological control of northern jointvetch in rice with an endemic fungal disease. *Weed Science* 21 (4): 303–307.

DANIELSON, L. L., ENNIS, W. B., FRANK, P. A., GENTNER, W. A., HAUSER, E. W., KLINGMAN, D. L. and SMITH, R. J. (1972). *Guidelines for weed control.* (Agricultural Handbook No. 447). pp. 171. USDA, Washington.

DE DATTA, S. K. (1972). Chemical weed control in tropical rice in Asia. *PANS* 18 (4): 433–440.

DE DATTA, S. K. and BERNASOR, P. C. (1973). Chemical weed control in broadcast-seeded flooded tropical rice. *Weed Research* 13 (4): 351–354.

DE DATTA, S. K. and LACSINA, R. Q. (1972). Weed control in flooded rice in tropical Asia. *Proceedings 11th British Weed Control Conference* 2: 472–478.

DE DATTA, S. K. and LACSINA, R. Q. (1974). Herbicides for the control of perennial sedge *Scirpus maritimus* L. in flooded tropical rice. *PANS* 20 (1): 68–75.

DE DATTA, S. K., LACSINA, R. Q. and SEAMAN, D. E. (1971). Phenoxy acid herbicides for barnyard grass control in transplanted rice. *Weed Science* 19 (3): 203–206.

GREEN, D. H. and EBNER, L. (1972). A new selective herbicide for rice, S-(2-methyl-1-piperidyl-carbonylmethyl)0,0-di-N-propyl dithiophosphate, for use alone or in mixtures. *Proceedings 11th British Weed Control Conference* 2: 822–829.

KASASIAN, L. (1971). *Weed control in the tropics.* pp. 307. Leonard Hill, London.

KIMURA, I., KAWANO, K., SADOHARA, H. and YOSHIDA, Y. (1975). Synergism in benthiocarb-propanil combination. *Abstracts 5th Asian-Pacific Weed Science Society Conference,* 1975. p. 63.

MATSUNAKA, S. (1975). Tadpole Shrimp: a biological tool of weed control in transplanted rice fields. *Abstracts 5th Asian-Pacific Weed Science Society Conference,* 1975. p. 11.

MITTRA, M. K. and PIERIS, J. W. L. (1968). Paraquat as an aid to paddy cultivation. *Proceedings 9th British Weed Control Conference* 2: 668–674.

MUKHERJI, S. K. (1968). Chemical control of algae in West Bengal paddy fields. *World Crops* 20 (1): 54–55.

OKAFOR, L. I. and DE DATTA, S. K. (1976). Chemical control of perennial nutsedge (*Cyperus rotundus* L.) in tropical upland rice. *Weed Research* 16 (1): 1–5.

OZAWA, K. (1975). (C–I) Herbicides: Rice. *Japan Pesticide Information* 25: 5–10.

PARKER, C. and DEAN, M. L. (1976). Control of wild rice in rice. *Pesticide Science* 7.

RAI, B. K. (1973). The red rice problem in Guyana. *PANS* 19 (4): 557–559.

SHINDO, N. (1975). The development of herbicides in Japan. *Japan Pesticide Information* 22: 5–11.

SINGH, G. and SAINI, S. SINGH (1960). Three new purple leaved strains to help you wipe out wild rice from your rice fields. *Indian Farming* 10: 6.

SMITH, R. J. (1967). Weed control in rice in the United States. *Institute for Technical Interchange, Weed control basic to agricultural development. First Asian-Pacific Weed Control Interchange,* Honolulu, University of Hawaii East-west Center, 1969. pp. 67–73.

SMITH, R. J. (1971). Red rice control in rice. *Proceedings 24th Annual Meeting Southern Weed Science Society,* 1971, p. 163.

SMITH, R. J. and SHAW, W. C. (1966). *Weeds and their control in rice production.* (Agricultural Handbook No. 292). USDA, Washington, pp. 64.

SOUTHERN WEED SCIENCE SOCIETY (1973). *Research Reports 26th Annual Meeting Southern Weed Science Society,* 1973. pp. 85–91.

SOUTHERN WEED SCIENCE SOCIETY (1975). *Research Reports 28th Annual Meeting Southern Weed Science Society,* 1975. pp. 98—102.

TAIWAN (1968). *Weeds found on cultivated land in Taiwan.* Department of Agronomy, College of Agriculture, National Taiwan University, Vol. 1, pp. 505.

TAKEMATSU, T., KONNAI, M. and ICHIZEN, N. (1975). Herbicidal activities against weeds and phytotoxicity on rice varieties of benthiocarb. *Abstracts 5th Asian-Pacific Weed Science Society Conference,* 1975. p. 62.

UMEDA, Y., HIGASHIKAWA, K. and FUKAZAWA, N. (1975). Thiochlormethyl, an interesting compound for herbicide mixture with broad spectrum weed control in rice culture. *Abstracts 5th Asian-Pacific Weed Science Society Conference,* 1975. p. 64.

DISEASES

FUNGAL DISEASES

In this section the most important fungal diseases of rice are discussed. Since the last edition of the manual there have been some changes in taxonomy. However, the most important changes are those concerned with disease control. The last few years have seen the banning of organomercurial pesticides. This has led to many changes in recommendations. The systemic fungicides have come into wide commercial use and antibiotics are now used on a field scale and show considerable promise for the future. Recommendations vary considerably according to varietal and environmental factors and are continually being revised. Readers are therefore advised to consult local extension officers to obtain current local recommendations.

Pyricularia oryzae
Blast

Distribution

Blast is the most widespread disease of rice. Over 70 countries have reported its presence (CMI Distribution Map 51, Fig. 17). The disease is present in almost every area where rice is grown commercially.

Symptoms

The disease affects all aerial parts of the plant. It is most conspicuous when it attacks the leaf blade and the neck, but it is also found on the leaf sheath, rachis, the joints of the culm, and the glumes (plate 5).

On the leaves the spots, first appearing as minute brown specks, typically enlarge to become spindle-shaped, pointed at both ends, several centimetres long and up to about 1 cm broad (Figs. 16 and 18). The centre is greenish-grey, often with a water-soaked appearance, later drying out to straw colour, and the margin is brown. The size, colour and shape of the spots vary considerably in different conditions and depend on the resistance of the variety affected. On resistant varieties or under unfavourable conditions lesions are small, brown, and narrow, and sometimes may not develop beyond the speck stage. Typical large, spindle-shaped spots occur on susceptible varieties and when conditions are favourable for fungus development. When several spots appear on a leaf they may coalesce and the whole leaf then withers. Leaf spotting is usually most severe on seedlings in the nursery, where severely affected plants may be completely killed.

Damage is most serious when the neck region of the flowering stem is attacked. The blast lesions appear at or near the uppermost node, where greyish-brown necrotic areas are formed, later becoming black; the tissues of the stem rot, and

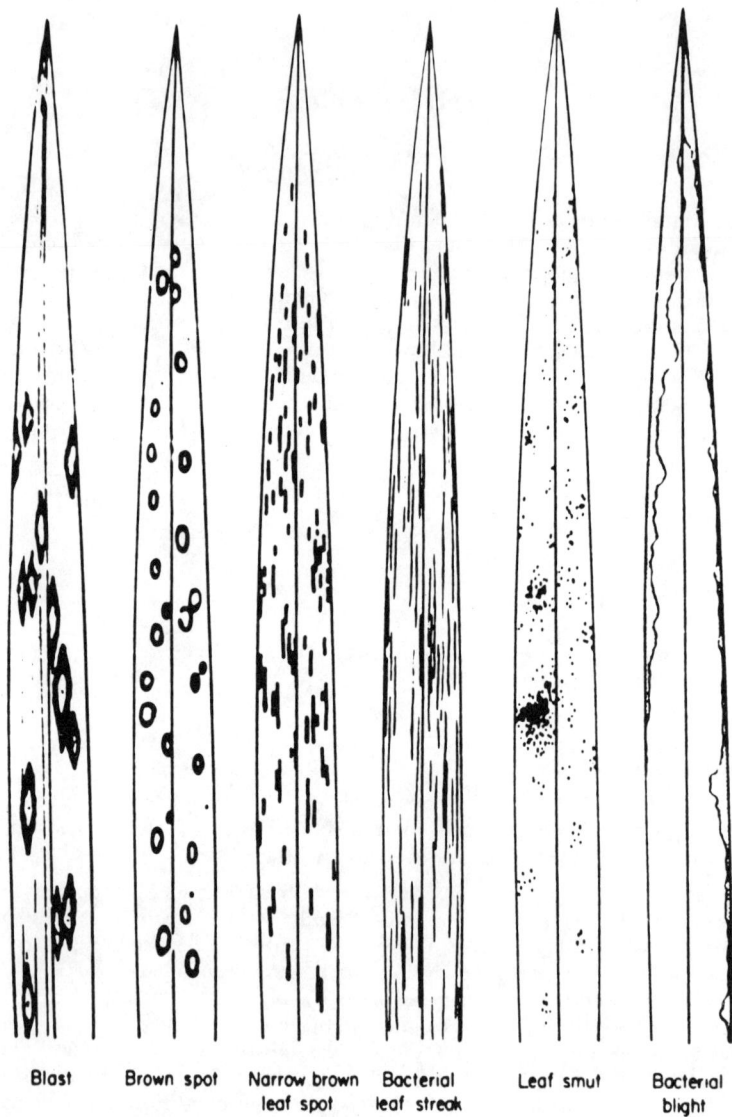

Fig. 16. Typical lesions of common leaf diseases of rice. (IRRI)

Blast   Brown spot   Narrow brown leaf spot   Bacterial leaf streak   Leaf smut   Bacterial blight

the panicle falls over, giving the typical broken neck phase of the disease (Fig. 19). The neck tissues are most susceptible when the ear is just emerging; if attack occurs at that time the grains will be empty. Later attack gives partly filled grains but the kernels are chalky and brittle, or may be green, and are useless.

The lower nodes of the stem may also be affected, turning black and rotting; the culm can then be pulled out and the blackened areas seen. Brown to black spots may also appear on the rachis, particularly where it branches, and on the glumes.

Blast is generally considered to be the principal disease of rice owing to its wide distribution and its destructiveness. Early infection often results in death of the affected plants. Panicle infections tend to reduce yield. In 1960, in Japan, blast accounted for 24.8% of total yield losses due to all factors. In 1953, a bad year, it was estimated that yields were lowered by 800,000 t. In epidemic areas of the Philippines 50% yield losses have been caused on thousands of hectares.

In Japan yield losses due to blast over the last 10 years have fluctuated. Generally the importance of blast has decreased, possibly because of improved farming techniques and the application of fungicides.

Blast is of major importance in all rice growing areas of Nigeria.

Taxonomy

Pyricularia is frequently referred to as *Piricularia* in the literature.

Development and spread

The fungus can be seen on leaf spots and other affected areas as a greyish or grey-green mouldy coating. Individual conidia are almost hyaline to pale olive, obclavate or pear-shaped, tapering at the apex, with two septa. They are borne on simple conidiophores and are readily detached and disseminated by air movement.

P. oryzae consists of many physiologic races which differ in their ability to infect rice varieties. Seventeen races are reported from Japan, 10 from Korea, 19 from Taiwan, 30 from India, 14 from Colombia and 25 from the USA. Recently 81 races have been identified in the Philippines by pathologists at the International Rice Research Institute. These investigations into the physiologic races of *P. oryzae* have been carried out independently in different countries using different sets of differential varieties; direct comparison of the races between countries is therefore not possible. Nevertheless, it seems that the prevailing races in Japan and Taiwan are quite distinct from those in the Philippines. Plant pathologists have been working towards the acceptance of an international set of differentials for use in identifying races, and as a result of a cooperative project between USA and Japan, 8 varieties have recently been suggested for international race group differentiation.

Work at IRRI has shown that the fungus is apparently heterocaryotic as many races of the fungus can arise from a single conidium, however this theory is not supported by Japanese pathologists.

Commonwealth Mycological Institute Map No. 51, issued 1.4.1981.

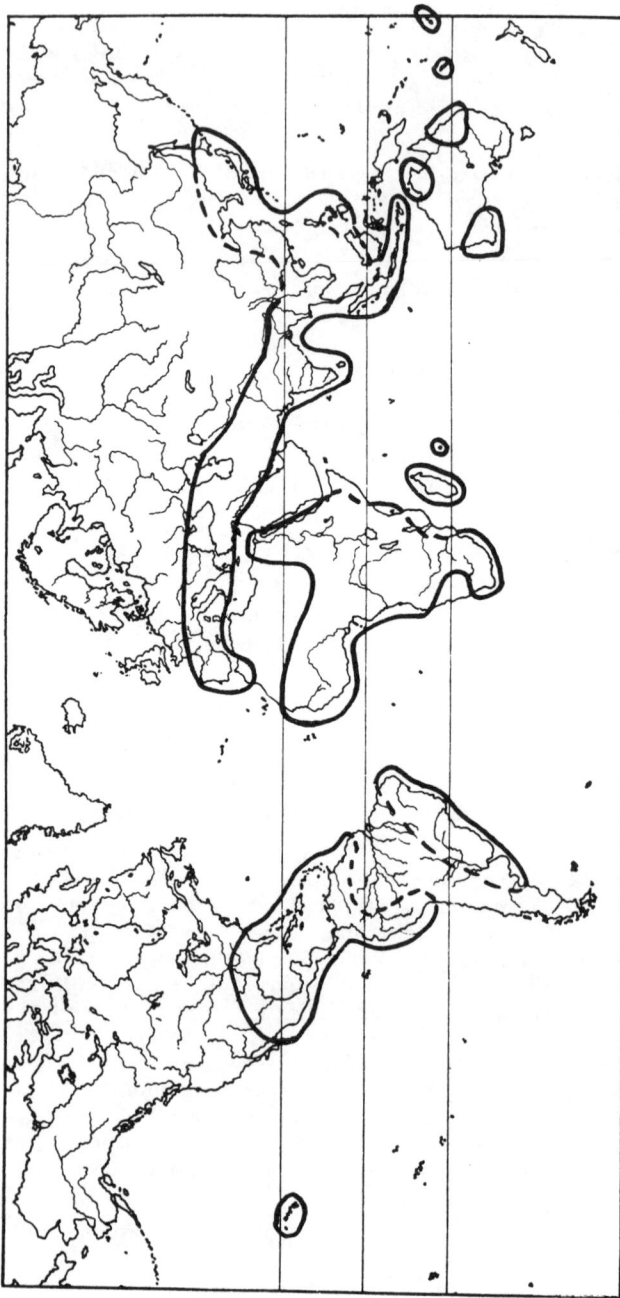

Fig. 17. Distribution of blast, *Pyricularia oryzae* on rice and other Graminae.

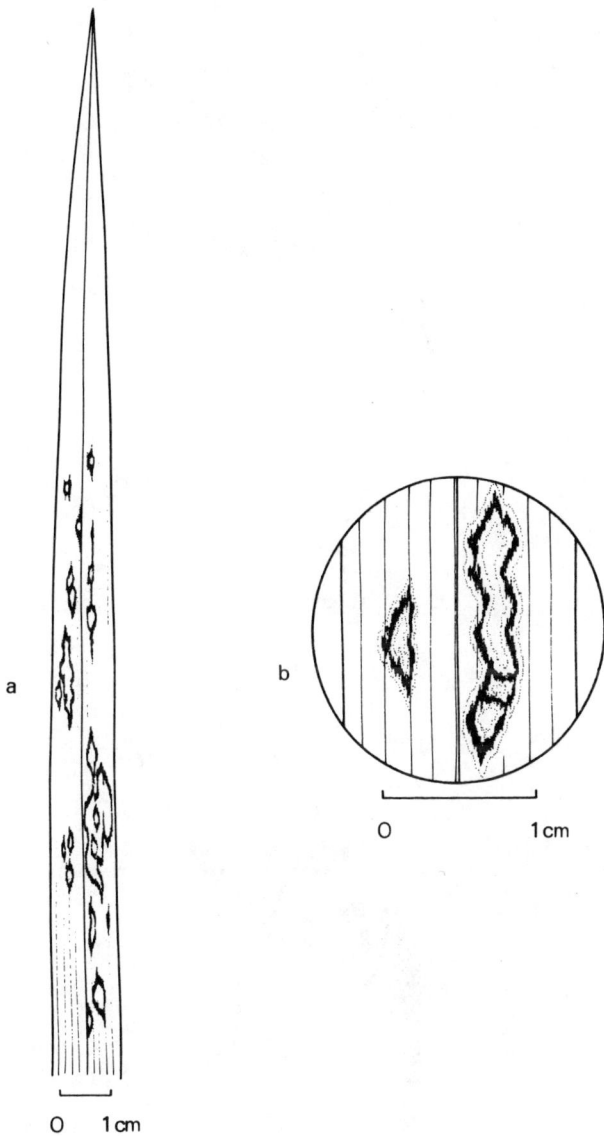

Fig. 18. Rice blast, *Pyricularia oryzae* leaf symptoms; entire leaf showing lesions (a) and lesion magnified showing distinct dark margin and typical spindle shape (b). (H.R.B.)

a

(diseased node)

(diseased node)

O   2cm

b

5mm

O

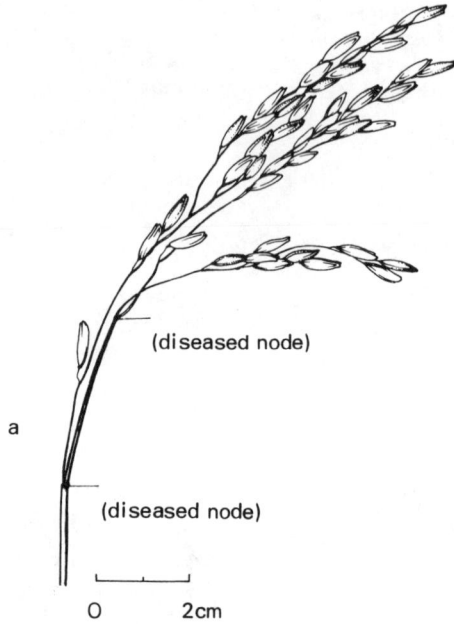

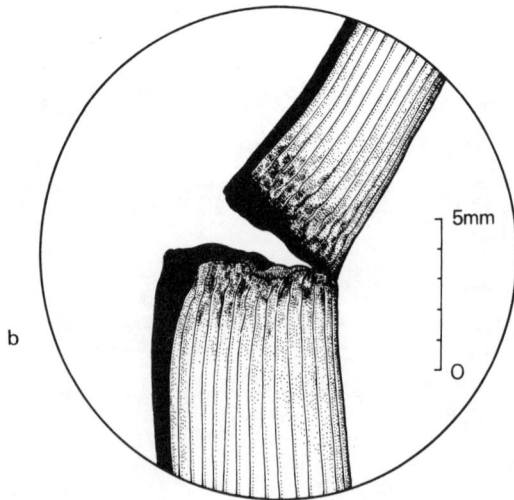

Fig. 19. Rice blast, *Pyricularia oryzae* neck blast symptoms;
panicle showing lesions on the uppermost node (a) and
node magnified to show 'broken neck' phase (b). (H.R.B.)

Conidia that have lodged on leaves or necks require free water (dew, etc.) for germination before penetrating directly into the host tissue. It takes only about 4—5 days for a spore to develop a visible lesion after infection, and the cycle of the disease repeats very quickly. High humidity favours the development of lesions; in low humidity brown margins develop on the lesions and their expansion is limited.

Numerous conidia are produced under humid conditions, usually during the night. A typical lesion produces as many as 4—5,000 conidia each night and continues to do so for 10—14 days under laboratory conditions. In extreme cases 15,000,000 conidia may be deposited on a square metre of rice field in a day.

Liberated conidia are carried upward by convection currents to the top of the plant layer, and from there are further disseminated by wind.

Experiments in Japan have established that conidia may be lifted to 25 m above the ground and carried horizontally for 20 km. Conidia have also been trapped on slides in aeroplanes flying at over 2,100 m.

Investigations on the influence of weather on spore survival have shown that spores kept in the hot sun for one day and returned to moist conditions do not develop but some of those kept in shady surroundings can survive.

In the tropics the disease is most destructive in the seedling stage and as neck rot after flowering. Very little blast is noticed between transplanting and flowering. In temperate countries, however, it may also be very destructive during the tillering stages. This may be due to differences in weather conditions, particularly moisture. The development of blast is certainly greatly affected by weather conditions; one important factor in the tropics is night temperature, low night temperature being the critical factor predisposing rice to infection. Varieties differ in their temperature relations. Low night temperatures can cause blast to develop on varieties which are normally resistant.

Investigations are being conducted at IRRI to discover why upland rice is so much more susceptible to blast than lowland rice. One line of investigation currently being followed is the importance of dew formation. Experiments have shown that at 26°C the fungus requires 12 h of wetness to develop and cause the disease. Owing to differences in edaphic conditions dew often does not last for 12 h in lowland regions whereas in upland regions it does. This may account for the difference in susceptibility.

Other predisposing factors include: application of large amounts of nitrogenous fertilizers; peat, volcanic ash or sandy soil; dry soil and cold irrigation water.

Control

Methods of use in controlling blast can be summarised as the use of resistant varieties, cultural methods, and the application of fungicides; a combination of these methods would normally be used.

Cultural

Time of planting can influence blast development. In Japan less blast occurs in early plantings.

It is well established that application of excessive amounts of nitrogenous fertilizers make the rice plant more susceptible to blast, and must therefore be avoided. In tropical areas comparatively little manuring of rice has been done in the past and this factor has been of slight importance, except sometimes in nurseries. There is, however, a trend towards a greater use of fertilizers in tropical areas, and avoidance of imbalance is important.

In the nursery, close planting should be avoided as this encourages development of the disease. Dry nurseries also encourage blast and wet nurseries are for this reason preferable when attack is anticipated. In Thailand Ou found that planting nurseries in small, narrow beds rather than large blocks reduced infection, as no areas of high humidity occurred in such narrow beds.

Resistant varieties

An enormous amount of work has been carried out, and is still in progress, on developing varieties with resistance to blast. It is not within the scope of this manual to discuss this work comprehensively.

Because of the presence of different pathogenic races in different localities, varieties resistant in one locality may not be resistant in other places. Tests and selections, therefore, should be made in each locality. Varieties of the *indica* group are more resistant in temperate countries and the *japonica* varieties tend to be more resistant in the tropics.

On occasion, varieties resistant to leaf blast have nevertheless seemed to be susceptible to neck rot. Studies at IRRI have shown that this is due to different races attacking the plants at different stages. To the same race, resistance or susceptibility is the same for both stages.

Variation between individuals of the one variety has been observed. This may be due to mechanical mixture of seeds, out-crossing and segregation, or other genetic changes.

The use of resistant varieties represents one of the principal methods of controlling blast. Advice should be sought from local extension officers in choosing the best variety for a particular area.

Chemical

Chemical control is widely used in the control of blight and has helped to diminish the importance of this disease in recent years.

Chemicals are used for seed treatment and as field sprays. Organomercurial compounds used to be commonly used but these are now banned in most countries owing to their toxicity.

In tropical areas, with lower yields per hectare, more extensive agriculture, lack of capital, often high cost of imported fungicides, and sometimes less knowledgeable farmers, the use of chemical control measures is not so highly developed as in Japan; with the use of unimproved varieties and without fertilizers, spraying with fungicides would in any case rarely be economic. With a general improvement in agronomic practices fungicidal treatment may become possible in some areas, and investigations into the fungicidal treatment of rice in tropical areas has begun and some recommendations have been made.

Seed treatment with benomyl 20% a.i. + thiram 20% a.i. has replaced mercuric seed treatments and is active against *Pyricularia, Gibberella fujikuroi* and *Cochliobolus*. The seed may be coated with the fungicide mixture at a rate of 0.5% of the seed weight, or slurried in a 2% solution of the fungicide mixture for 30 min or dipped in a 0.2% solution for 6—12 h. Standard boxes of seedlings for transplanting may be drenched with benomyl. Benomyl can also be used as a field spray at 300 g a.i./ha in 1,000 l of water at flowering with a second application 12 days later.

Antibiotics are now commonly used for blast control. These include blasticidin -S and kasugamycin. Edifenphos and IBP are effective fungicides for blast control. IBP is applied at 830—1,250 ml 48% e.c./1,000 l water/ha. One or two sprays may be needed during the head season. Alternatively IBP 17% granules applied at 30—45 kg/ha 7—10 days before the estimated date of the first blast lesions is recommended for leaf blast control.

Edifenphos is applied at 350—800 g a.i./1,000 l/ha, depending on the formulation. One or two applications are made to wet paddy in the nursery and two to three applications are made in the field. This fungicide is also effective against ear blight and sheath blight.

Pentachlorobenzyl alcohol 4% dust at 30 kg/ha or 50% w.p. at 1 kg/1,000 l water (with a spreader) applied before development of the disease is recommended as a protective treatment. In severe cases this can be repeated 7 to 10 days after the first application. For control of ear blast, neck blast and panicle blast 4% dust at 40 kg/ha or 50% w.p. at 1.2 to 1.5 kg/1,200—1,500 l water (with a spreader) should be applied before the production of ears and in severe cases repeated when ears are produced.

Kasugamycin gives best control when applied in the early stages of blast development. For control of leaf blast application should be made before or just after the occurence of the disease and repeated 5—7 days later. For control of neck rot and node blast application should be made just before heading and repeated 2 to 3 times at 7 to 10 day intervals. Recommended rates for the 2% liquid and 2% w.p. formulations are 1 l or 1 kg/ha in 300—1,000 l of water. The volume of water depends on the type of sprayer used. The 0.2% dust formulation is recommended at 30—40 kg/ha.

Blasticidin-S is recommended at rates of about 10 g/1,000 l/ha.

Cochliobolus miyabeanus
Brown spot

Distribution

Brown spot is a wide-spread rice disease occurring in all rice-growing countries; Asia, America and Africa (see CMI Distribution Map 92, Fig. 20).

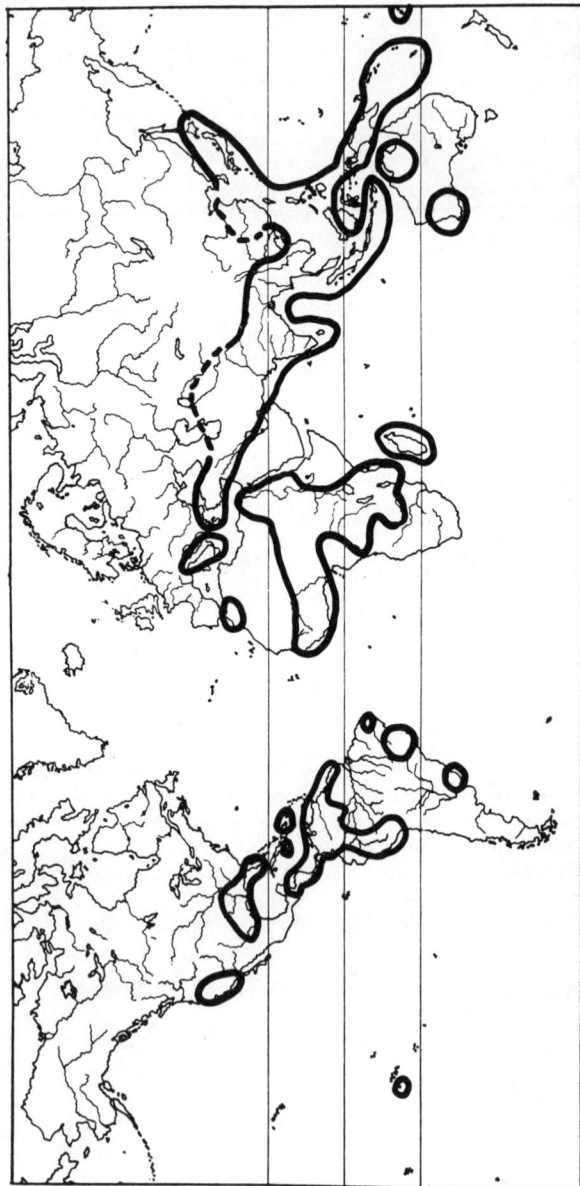

Commonwealth Mycological Institute Map No. 92, issued 1.10.1966.

Fig. 20. Distribution of brown spot, *Cochliobolus miyabeanus* on rice and other *Oryza* species.

Taxonomy

The fungus causing brown spot is *Cochliobolus miyabeanus* (Sphaeriales). The imperfect stage is *Drechslera oryzae*.

Symptoms

The fungus causing brown spot is capable of affecting the coleoptile, leaf blades, leaf sheaths, and glumes but is most commonly seen on the leaves. Spots first appear as minute brown dots, later becoming ellipsoidal or oval to circular with, when fully developed, a light brown, fawn, or grey centre and a dark or reddish-brown margin (Figs. 16 and 21). Seedlings are often heavily attacked, with very numerous spots about 2.5 mm in diameter; in such cases leaves may dry out and die. Yellowing leaves and dieback may also be a symptom. Badly affected nurseries can often be recognised from a distance by their brownish, scorched appearance. Seedling mortality is not usually high. Spots also appear on leaves of older plants, on which they become larger than on seedlings. They may be up to about 1 mm in length but are usually smaller; the larger spots are typical of the more susceptible varieties (Plate 4).

On the coleoptile the spots are brown and small. On the glumes the disease appears as dark brown or black oval spots (Fig. 22) or, when attack is more severe, the whole surface of the grain may be blackened, and the seeds are shrivelled and discoloured.

The disease is seed-borne and seedlings grown from infested seed become blighted. The disease can cause heavy losses. The leaf spotting stage, which can reach epidemic proportions, is particularly important. Brown spot was considered to be a major contributory factor of the 1942 Bengal famine when losses of 50–90% were reported. In the Punjab losses of grain weight ranging from 4.6–29% have been reported.

C. miyabeanus is the most important causal agent of ear blighting. This disease is caused by *C. miyabeanus*, *Cercospora oryzae*, *Fusarium nivale* (*Micronectriella nivalis*), *Helminthosporium sigmoideum* (*Leptosphaeria salvinii*) and *H. sigmoideum* var. *irregulare*. Ear blighting is rampant in Formosa, Thailand, India and other Asian countries. The symptoms are very similar to those of rice blast with dark brown discolouration of the uppermost internodes, neck nodes and panicles with poor grain.

Development and spread

Spores germinate, under favourable conditions, within a few hours and form appressoria on the surface of leaves etc., these form hyphae which penetrate directly through the cuticle and epidermis, or through the stomata. The first sign of infection may appear within 24 h and within a few days spores may be produced, so that secondary spread of the disease can be rapid; it is this quick secondary spread which, under conditions favourable to the fungus, can be responsible for severe outbreaks.

The fungus may survive for two to three years in infected plant parts, particularly the seeds, and it is from infected seeds that the disease often starts. In India it

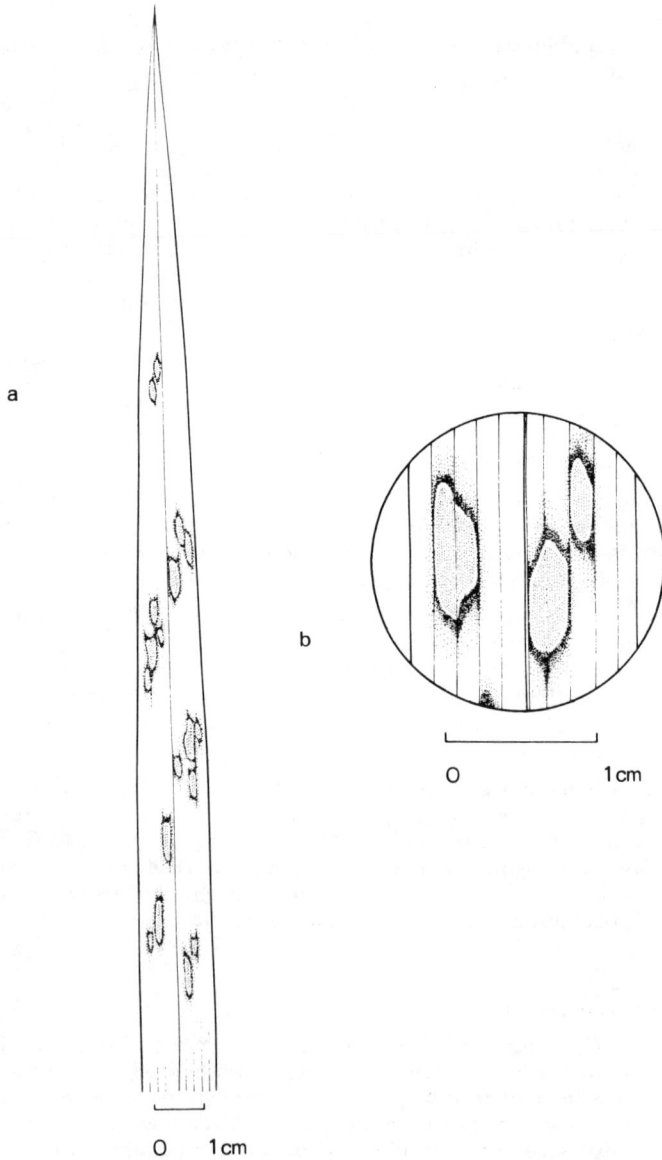

Fig. 21. Brown spot, *Cochliobolus miyabeanus* leaf symptoms; entire leaf showing ellipsoidal lesions (a) and magnified lesions showing dark margin and lighter centre region (b). (H.R.B.)

O   2cm

Fig. 22. Brown spot, *Cochliobolus miyabeanus*
on panicle showing dark brown and black oval
spots on the glumes. (H.R.B.)

has been found that conidia are present in the air near rice fields throughout the
off season and in many parts of the tropics such conidia could form another
source of primary infection. Several weed grasses can be infected by *C. miyabeanus* artificially and these grasses (e.g. *Echinochloa colonum* and *Leersia
hexandra*) growing around rice fields have been reported as natural hosts of
the fungus. It seems likely, however, that infection of such grasses is of little
importance in the disease cycle. The fungus may also perennate in stubble, as
reported from India. Airborne conidia produced on young seedlings cause
secondary infection on older leaves and later on the grains.

Temperature, moisture and light all influence the development of brown spot.
There are differing reports on the optimum temperature for infection but this
appears to be in the region of 25°C. Free water on the leaf surface favours
infection. Shading appears to increase infection and dry soil has a deleterious
effect.

It has, however, generally been noted that the main factor governing disease
incidence is the physiological condition of the rice plant, and this is mainly
governed by soil conditions. So, in Japan, plants grown in sandy, peaty, or thin
soil, with poor drainage and low nutrient status, are most susceptible to brown
spot.

There is a positive correlation between the incidence of the physiological disease 'akiochi' (see chapter on physiological diseases) and that of brown spot.

In India, crops grown under well-balanced nutrient conditions are usually not found to be severely attacked by brown spot which is generally confined to comparatively poor, leached soils, when water is lacking.

In Malaya, where the disease is most serious in the seedling stage, damage is most severe in dry nurseries planted in poor soil without added nutrients. Brown leaf spot occurs in almost all areas of Nigeria where upland rice is grown. Serious infections have been found associated with heavy shading, potash deficiency and soil salinity. Deficiency of manganese, iron or magnesium also increases the susceptibility to brown spot.

Control

Cultural

The incidence of brown spot is closely correlated with mineral nutrition; careful use of fertilizer can do much to prevent the disease. Application of open-hearth furnace slag containing silicon, iron, magnesium, manganese and phosphate is effective in controlling the disease.

Application of fertilizer in 2 or 3 split applications during the season prevents the nutritional decline at the end of the growth period which is often associated with brown spot. Deep ploughing and the use of compost and potassium fertilizer are also recommended.

In the tropics other rice plants or alternate hosts provide sources of inoculum, so control by crop rotation is not practical because the mycelia and conidia can survive for so long in the field.

Resistant varieties

Resistant varieties offer the best means of control.

As resistance to brown spot is correlated to resistance to H_2S, resistance can be measured by testing varieties in dilute H_2S solution. There is a tendency for *indica* varieties to be more resistant than *japonica* varieties which correlates with their respective ability to absorb nutrients.

A considerable amount of work has been done on selecting and breeding resistant varieties. Much of this work has been done in India. Of 490 varieties tested Ch13, Ch45, T-141, T-498-2A, Co20, BAM10, T998m, T-2112, T-2118 and T-960 were found to be resistant. In a more recent trial the following were found to be resistant: from Japan, Kusaful, Kinmaze, Hoyoku and Japan 11; from America, varieties Rexoro and C5309; from Taiwan, varieties, Chainung 242, Chainan 8, Taichung Native 1, Tainan 1 and Tainan 3; from Hong Kong, Fa Yiu Tsai and from IRRI, IRRI 29, 31, 44, 47, 48, 49 and 50. Work has been carried out in the USA to breed varieties resistant to both brown spot and blast.

Chemical

As the brown spot fungus is seedborne, seed treatment with fungicides is recommended in several countries.

In laboratory tests mancozeb seed treatment (1,000 ppm for 6 h) was more effective than the previously used mercurial treatments. 20% a.i. benomyl + 20% a.i. thiram is an effective seed treatment against brown spot. The seed may be coated with the fungicide at a rate of 0.5% of the seed weight by shaking the pesticide and seed together in a sack. Alternatively the seed can be slurried in a 2% solution of benomyl + thiram for 30 min or dipped in a 0.2% solution for 6—12 h.

Hot water treatment is also effective but is not usually practical at the ordinary farmer's level.

Preliminary investigations in Ibadan, Nigeria have shown that mancozeb 80% a.i. (1 g/l/10m^2), blasticidin-S 4% a.i. (1 g/l/10m^2) and thiram 65% a.i. (2 g/l/10m^2) have good potential as foliar sprays for the control of brown spot.

Various other evaluations of field spraying to prevent secondary infection have been made but their practical usefulness is doubtful. Spraying with anilazine has been shown to be efficacious against ear blighting.

The antibiotic polyoxin A is very active against this disease.

Magnaporthe salvinii
Helminthosporium signoideum var. irregulare
Stem rot, irregular stem rot

Distribution

The organism causing stem rot was first recorded in Italy in 1876 and has long been known in many other areas. It has a wide distribution in Europe, Asia and the Americas and has been reported from several African countries (see CMI Map 448). The organisms causing stem rot often occur in the same field.

Taxonomy

The causal agent of stem rot is *Magnaporthe salvinii*. The imperfect (conidial) state of this organism is *Nakataea sigmoidea*. In Japan stem rot is usually referred to by its imperfect state name. The sclerotial state is *Sclerotium oryzae*.

H. sigmoideum var. *irregulare* is the causal agent of irregular stem rot. No perfect state of this organism is known.

Symptoms

The symptoms of both diseases are very similar and as the two frequently occur together they will be dealt with in the same section. The main difference is that the sclerotia of *H. sigmoideum* var. *irregulare* are irregular with a rough surface and 90—119 x 268—342 μm in size, whereas those of *M. salvinii* are spherical with a smooth surface and usually 230—270 μm in diameter. The conidial stages are also morphologically distinct.

Infection usually takes place through the leaf sheaths, although it may also be through the roots or the stem base. It often arises from sclerotia which are freed from the soil or plant debris during ploughing and disseminated by irrigation water. A brown or black, oval necrotic area develops on the outermost leaf sheath at about the water line. The rot spreads to involve the whole sheath and the leaf blade yellows and withers. The fungus penetrates inwards, killing the leaf sheaths and causing the leaf blades to wither successively as it progresses. Finally it attacks the culm, where one or more internodes may be infected; rotting and breaking of the affected stem causes lodging.

If the infected stem internodes are split open the dark greyish mycelium is observed within the hollow stem. Black sclerotia can be seen scattered all over the inner surface (Plate 4).

Sclerotia may also be found on the surface of the leaf sheaths as well as inside them, where they often appear as rows of black dots; and these bodies also occur on other diseased plant parts. They constitute the easiest feature of the disease to recognise and their occurrence is therefore a useful diagnostic character.

Another symptom sometimes associated with stem rot is the very late production of numerous small tillers from the base of the affected plant; by the time they appear the original tillers, if surviving, are already mature.

When young plants are attacked they may be killed; but more commonly the disease appears at a fairly late stage of tillering and the plant survives; little grain is produced, however, because of lodging. The abnormal late tillers produce nothing and any ears which succeed in emerging often have a high proportion of light grains.

Stem rot is one of the more serious rice diseases as far as yield losses are concerned and can be second only to blast in its severity. In Japan in recent years the annual loss in yield was estimated to be 16,000—35,000 t. High yield losses are reported from most years. In epidemic years losses of up to 80% have been reported. In a large scale survey in the Philippines 67% of the tillers examined were infected with one of the stem rot organisms. As the organisms rely on wounds as an entry site it is probable that the disease severity depends on factors such as lodging and other physical damage rather than simply the presence of the causal organism.

Development and spread

Numerous sclerotia are left in the field after harvest, and may survive for several months or a year. During ploughing and cultivation for the next crop some of the viable sclerotia float to the surface and come in contact with a leaf sheath or stem of a rice plant, where infection takes place through a wound. Under certain conditions infection may also take place through the roots or the stem base below soil level. Conidia may be produced on diseased leaf sheaths and can cause secondary infection when carried by wind to a new infection site. Irrigation water carries sclerotia from field to field and so spreads the disease with the flow of the water.

A major factor affecting the severity of attack is the water supply. In Japan, stem rot is only found in ill-drained fields and is intensified if the water is kept deep at tillering. In India it has been noted that incidence of stem rot is highest in heavy, badly drained soil in low-lying areas and when the water remains stagnant for long periods.

The disease is also influenced by the nutrition of the rice plant, excess nitrogen favouring attack, especially if applied early. Conversely it has been shown that high rates of potash fertilizer reduce infection, particularly if the rate of postash is high in relation to nitrogen.

In Italy, although M. *salvinii* is widespread and can usually be found on a high proportion of rice plants, it causes little damage unless the infected plants are damaged e.g. by unfavourable weather causing breakage of stems. Studies at IRRI confirm that the fungus is mainly a wound parasite, entering the plant through damage caused by insects, lodging, etc. In the absence of wounds the fungus lives as a saprophyte on dead or moribund tissues.

Control

The control measures against both stem rots are the same.

Resistant varieties seem to offer the most effective means of control. However some varieties are more resistant or more susceptible to one fungus than to the other.

Several workers have screened rice varieties for stem rot resistance. Numerous varieties including Gimbozu, Shinriki, Sembon, IRRI-2, IRRI-25 and IRRI-34, show considerable resistance.

Fungicides are not widely used for control of stem rot. IBP, edifenphos, benomyl and blasticidin-S will all control the disease.

Cultural methods of control include proper water management, where this is possible and avoiding too deep water, especially at tillering. Stagnant water is also to be avoided. Draining off the water completely and allowing the soil to dry out for a time has also been found effective but this is seldom possible in the tropics, where rice fields are mostly dependent for water on unpredictable rainfall.

As the fungus can perennate in plant remains, burning off the stubble and straw left in the field after harvest is an added precaution. But even under the best conditions burning the stubble sufficiently thoroughly to kill all the sclerotia present is difficult.

When fertilizers are used it is important to avoid an excess of nitrogen, as this renders the rice more susceptible.

Corticium sasakii
Sheath blight

Distribution

Sheath blight was first reported in Japan in 1910 and now occurs in most rice growing countries.

Damage is often serious and the disease is considered to be of major importance in several countries including Sri Lanka, China, Taiwan, Japan, Indonesia, Philippines and Vietnam. In Japan only blast affects a greater area. Sheath blight was first reported from Brazil in 1973. In Nigeria sheath blight is usually found on the variety OS6 planted under upland conditions but the incidence is not high enough to be of importance.

Taxonomy

There is considerable disagreement on the name of the causal organism of sheath blight. *Corticium sasakii* is one of the most commonly used names. Other names used include *Hypochnus sasakii* and *Pellicularia filamentosa*. The imperfect stage of the organism is *Rhizoctonia solani*. Recent taxonomic studies have indicated that the perfect stage should be called *Thanatephorus cucumeris*.

Symptoms

Plants are usually attacked around the tillering stage, when leaf sheaths become discoloured at or above water level. Large lesions are formed, at first greenish-grey and later becoming fawn or off-white with a brown or purplish-brown margin. The lesions are large, oblong or irregularly elongated, and appear on any part of the leaf sheath, sometimes extending on to the leaf blade (Fig. 23). Eventually the whole sheath rots and the leaf can easily be pulled off. Outer leaf sheaths are first affected, the causal fungus later extending to the inner sheaths. At a later stage, particularly in humid conditions, the sheath surrounding the ear may be infected, in which case the ear is prevented from emerging and expanding normally (Plate 5).

Seedlings may also be infected in the nursery if planted in infested soil, when the base of the plant is attacked and whole patches of seedlings may be killed.

Sclerotia of the causal fungus are sometimes conspicuous on diseased plant parts. They are white at first, later becoming brown or purplish-brown, variable in size up to 5 mm diameter. They are produced loosely on the surface of the leaf sheath and are roughly spherical or somewhat flattened. Sclerotia are also found between the leaf sheaths, when they are flattened and of irregular shape. Hyphae of the fungus can also sometimes be seen on the surface of the leaf, particularly in highly humid conditions, as very fine silvery threads. Sclerotia are easily detached so that they fall into the soil or water. Sclerotia can also infect *Rottboelia exaltata*, sorghum, sugarcane and maize.

Development and spread

The fungus is able to survive through the off season as sclerotia or mycelium in the soil or in rice plant remains. During preparation of the land for planting sclerotia are released from the soil into the water and can thus spread the disease.

The disease is favoured by high humidity, which occurs among the rice plants, particularly later in the season, when the plants have grown large and dense; it is for this reason that attack often occurs at this stage of growth. Primary infection frequently occurs just after transplanting, as floating sclerotia become

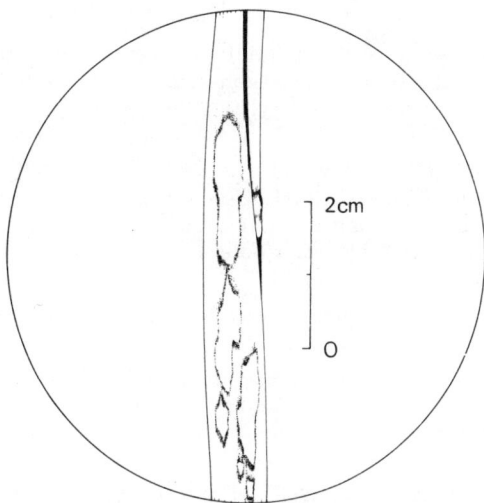

Fig. 23. Sheath blight, *Corticium* sp. infected plant
showing base of stem and leaf sheath with large
lesions. (H.R.B.)

attached to the tillers. Many-tillered plants trap a larger number of the floating
sclerotia than those with fewer tillers. Dense planting, by increasing humidity,
increases infection. Heavy applications of nitrogen are also conducive to attack,
perhaps because they induce leafy growth and therefore high humidity among
the plants.

In Japan sheath blight has become more serious and widespread since early
planting was widely adopted.

Control

Cultural

As the sheath blight fungi can survive in infected plant remains, these should be
destroyed when possible. Avoiding excess applications of nitrogen is essential.
In Japan it has been found that several grasses growing round rice fields are among
the numerous alternative hosts of *C. sasakii* and destruction of these weeds is
advocated.

Resistant varieties

Generally early varieties suffer more than late ones. In Japan, late maturing
varieties are more likely to escape the disease because of lower temperatures in
the autumn. *Indica* varieties are more resistant than *japonica* varieties.

In the Philippines, varieties considered to have some resistance include Zenith,
Kataktara DA-2, Kwang-fu 1, Pl 160641, and Mamoriaka.

Chemical

Many chemicals have been used in attempts to control sheath blight in Japan. The organoarsenicals e.g. ferric and calcium methane arsonates, methyl arsenic sulphide and methyl arsenic bis-(dodecyl sulphide), have been widely used with success.

Ferric methane arsonate is used in Japan to control sheath blight. Ferric methane arsonate does not show the same level of phytotoxicity as that found with methane arsonic acid. It has a preventative and curative effect but should not be applied during the period just after booting as it can adversely affect the developing seeds.

Organoarsenicals should be applied at least 10 days before heading to avoid phytotoxicity.

The current standard treatments in Japan are antibiotics. Such materials are preferable to the organoarsenicals which can be phytotoxic. The antibiotics validamycin and kasugamycin give good control. Polyoxin dust also gives a good control.

The two antibiotics, SAF-49 and MSF-711 and several new compounds including BAS-3271 F (N-cyclohexyl-2,5-dimethylfuran-3-carboxylamide) and IKF-214 (not published) have been tested for control of sheath blight but were less effective than arsenical fungicides.

Benomyl at 6.6 or 3.3 kg a.i./ha has been shown to be very effective against sheath blight. Hymexazol solution and methyl arsenic-bis-(dodecyl sulphide) are effective preventatives.

A new antibiotic, baridamycin, as a 3% solution and 0.3% dust has been shown to have excellent effects against sheath blight in field trials.

In many cases, especially in the tropics where yields are low, chemical control is unlikely to be economically worthwhile, but it may be justifiable in very heavy attacks. Otherwise control must largely depend on planting varieties which are not too susceptible, and on cultural control.

Corticium (Sclerotium) rolfsii

Seedling blight

This disease was first described from the Philippines and has since been found in several other rice-growing areas, including Malaysia, Thailand, USA and Malagasy. As the fungus causing the disease is very common on a wide range of host plants in most areas of the tropics (see CMI Distribution Map 311) the disease is likely to turn up in many other countries. In affected areas the occurrence of seedling blight is sporadic and, although individual nurseries may be severely affected, the overall effects of the disease are relatively unimportant. The disease tends to occur in upland rice seedbeds and in lowland seedbeds which dry up at times. Patches of seedlings are attacked in the nursery. Affected plants make slow growth and their leaves yellow and dry up. The cause of this is a rot of the base of the stem, which becomes dark brown. The rot is often fatal, and the result is the appearance of conspicuous patches of dead seedlings scattered through the nursery. Plants which survive until irrigation may recover as the fungal development is checked by watering.

The disease is caused by *Corticium rolfsii* (Basidiomycotina, Aphyllophorales) but as the perfect state is rare the imperfect state name *Sclerotium rolfsii* is often used. White strands of mycelium appear on the discoloured base of the stem, followed by the round, smooth sclerotia, at first white but later developing a typical light brown colour. The sclerotia are often described as being similar to mustard seed: they are about 0.5−1 mm in diameter and are easily removed from the diseased part and so fall to the soil, where they can survive for a long time and so perpetuate the disease. Sclerotia are dispersed by ploughing and in irrigation water.

Irrigation at the time of sowing reduces infestation and discourages further infection. The disease is more common during warm moist weather and is increased by a high organic matter content in the soil. Although the fungus can penetrate the plants in the absence of wounding, damage such as mechanical injury can make infection easier.

Deep ploughing to bury crop debris tends to reduce disease incidence. PCNB soil treatment has been reported to give control of seedling blight. Seed treatment with a mixture of benomyl (20% a.i.) plus thiram (20% a.i.) controlled both seed and soilborne seedling blight but slightly inhibited shoot growth.

Other seedling diseases

In recent years there have been changes in rice growing methods especially in countries like Japan where labour costs have risen steeply. In Japan over 50% of rice is now mechanically transplanted. Seeds are sown in standard seed boxes on an automated production-line. The boxes are then incubated at $28°C$ and nearly 100% r.h. After germination the seedlings are grown in polythene houses. The uniform turves are taken from the boxes for use on transplanting machines. The plants are about 10 cm high at this stage being much younger and smaller than seedlings for hand transplanting. Due to the increase in mechanical transplanting losses caused by *Rhizopus* sp. in the seed boxes, and *Pyricularia* sp. at an early stage in the field, have increased in importance. Mechanical transplanting is more likely to cause physical damage to the plants and thus likely to increase the chance of infection through wounds. Damping off has not been reported as being a problem in rice in the tropics. Cool weather tends to encourage damping off and injured seeds are more subject to infection. Cottony hyphal growth on the surface of the seeds in the nursery indicates the presence of damping off organisms. Severely affected seeds fail to germinate. Later infection results in damping off. Damping off is caused by several species of water moulds, the main ones being *Achlya* spp. and *Pythium* spp. Other organisms causing seedling diseases include *Fusarium* spp., *Helminthosporium oryzae*, *Rhizoctonia* spp. and *Corticium rolfsii*.

Various fungicide treatments have been evaluated for control of seedling diseases. Captafol and captan seed treatments have been reported to give good control of

Achlya klebsiana and *Pythium* spp., Hymexazol and benomyl soil treatment has given a good reduction of damping off in seedling boxes.

Thiram and chlorthalonil are other possible fungicides for damping off control. In the USA captan, maneb or thiram are usually used as seed-protectant fungicides. An emulsion containing 20% 2,3-dibromopropionitrile and 20% trichloronitroethylene is recommended for damping off control. This should be mixed at 10–15 ml/10 l of water and applied at 3–4 l/m^2. Application should be made over the whole field about a week before seeding.

Gibberella fujikuroi (Fusarium moniliforme)

Bakanae disease, foot rot

Distribution

Bakanae disease is widespread in tropical and temperate rice-growing areas in all parts of the world (see CMI Map 102).

Taxonomy

Bakanae is known by several other names including elongation disease, white stalk, palay lalake and man rice.

The disease is caused by a fungus, *Gibberella fujikuroi* (Ascomycotina: Hypocreales), sometimes mentioned in the literature under the name of its imperfect state, *Fusarium moniliforme.*

G. fujikuroi occurs on other important graminaceous crops, including maize, sorghum, sugarcane, and wheat, and on plants in many other families; but other hosts are probably not of importance in the disease cycle on rice.

Symptoms

In the nursery, affected seedlings are pale yellowish-green, thin, and abnormally elongated. They are usually scattered through the seed-bed and do not occur in definite patches. Many seedlings die but some survive transplanting and produce plants which at the tillering stage are often noticeably taller than normal plants and are also yellowish.

In older plants, which are taller than normal, the leaves dry up and turn brown, beginning with the oldest, the margins drying out first. Inside affected stems and lower nodes are discoloured brown. On the outside of dead leaf sheaths, just above water level, a white or pink bloom of fungus mycelium develops, later extending about 10 cm up the stem and developing into a pink encrustation. The root system is not affected and dead plants, when pulled up, tend to snap at the collar. Also typical of the disease is the production of adventitious roots from one or more nodes above water level.

Slightly affected plants which survive until maturity may produce panicles which appear slightly earlier than normal. They are small with few spikelets, and are sterile.

In some cases infected plants do not show the typical elongation but are instead severely stunted; this tends to occur at lower temperatures and when the soil is rather dry.

Development and spread

Bakanae disease is seedborne, infection taking place at flowering. The causal fungus can also persist in soil, but is short lived in soil in the tropics. There may also be carryover of the disease in straw and stubble. Bakanae disease is less common at low temperatures. The optimum temperature for growth of the causal organism is about 27–30°C whereas 35°C is the temperature most suitable for seedling growth and for infection.

Control

Losses due to bakanae have been estimated as being as high as 40% in Asia and almost complete crop failure in Australia. However the wide use of seed treatment means that the disease now no longer assumes its previous importance and is chiefly known for its role in the discovery of gibberellins. As mercurial seed dressings are now banned in most countries the effectiveness of other chemicals has been investigated.

Benomyl alone or in a mixture with thiram (20% a.i., of each) gives effective control. Dipping seeds in thiophanate-methyl solution (0.07–0.14% a.i.) for 1–6 h gives good control and cypendazole is also effective. Rice varieties vary considerably in their resistance to bakanae. Varieties native to temperate regions tend to be more resistant than those from the tropics.

Sphaerulina oryzina (Cercospora oryzae)
Narrow brown leaf spot, earblight

Distribution

Narrow brown leaf spot has a world-wide distribution, except for Europe (see CMI Distribution Map 71) and occurs commonly.

Taxonomy

The pathogen causing narrow brown leaf spot is *Sphaerulina oryzina* (Ascomycotina, Sphaeriales) which is the perfect state of *Cercospora oryzae* (Deuteromycotina, Hyphales).

Symptoms

The disease produces small, linear brown spots which are about 2–10 mm long and about 1mm wide and occur most commonly on leaves (Figs. 16 and 24). Usually the leaf spots are reddish brown with lighter coloured edges. The spots may be slightly wider, with a light narrow centre in susceptible varieties. Spots on the leaf blades may be very numerous and in severe cases cause the leaves to dry up from the tips downwards and to wither prematurely. Spots may also

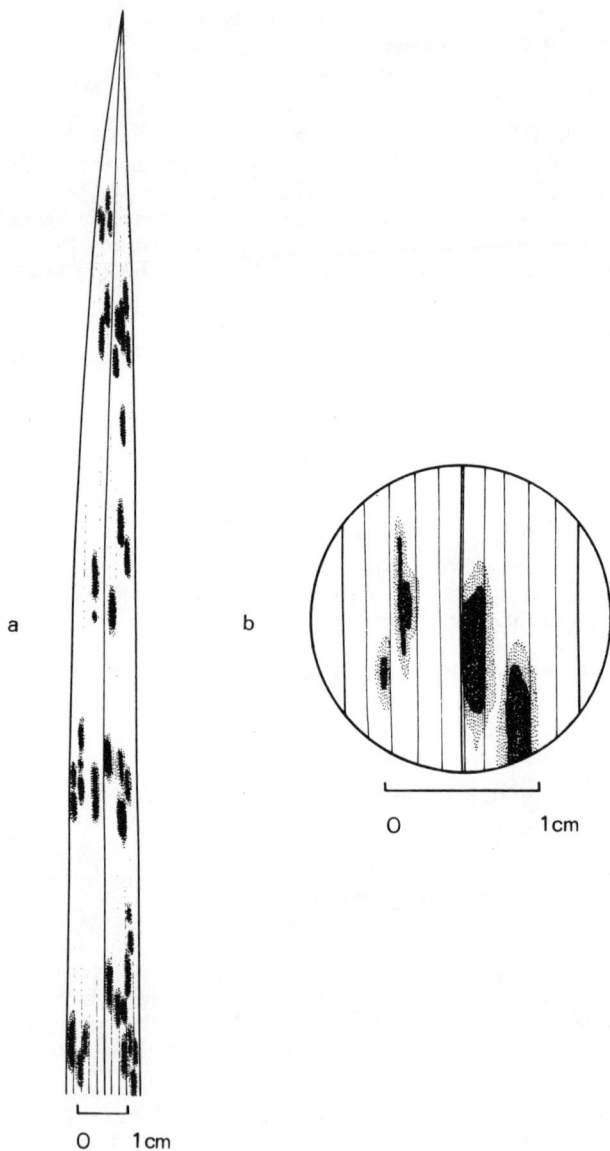

a   b

O   1cm

O   1cm

Fig. 24. Narrow brown leaf spot, *Cercospora oryzae;* entire
leaf showing small rectangular elongated lesion (a) and
magnified lesions showing their development in size and
intensity (b). (H.R.B.)

occur on leaf sheaths and on pedicels, glumes and grains where they cause ear-blight. There are several pathogenic races of the fungus.

Development and spread

Narrow brown leaf spot does not affect seedlings and usually appears quite late in the season. The fungus enters through the stomata and spreads longitudinally in the epidermal cells. Maximum severity usually occurs at about flowering time, however very susceptible varieties may become infected earlier. Damage caused by this disease is not usually serious as the attack develops in the late stage of growth. The disease is prevalent in the Gulf States of the USA where a great deal of attention was paid to it in the 1930s and 1940s when some commercial varieties were very susceptible. A 40% loss was reported in Surinam during 1953—4. A build-up of the disease in one area of Nigeria was reported for the 1963—4 dry season. Generally heavy damage is restricted to susceptible varieties.

Control

The use of control measures such as fungicidal spraying is not generally considered economic. Much work has been done on the development of resistant varieties, particularly in the USA. Some varieties were found to be resistant to certain races of the disease but susceptible to others. Resistant varieties include: Delrex, Sunbonnet, Selection 44C507 and Toro.

Entyloma oryzae

Leaf smut

Leaf smut is not a disease of great economic importance but it is widely distri-buted in rice-growing areas of the world (see CMI Distribution Map 451).

The smut causing the disease is *Entyloma oryzae* (Basidiomycotina: Ustilaginales). Minute, black, slightly raised spots are formed on the leaf blade and, less commonly, on the leaf sheath and stem (Figs. 16 and 25). If the spots are numerous the tip of the affected leaf withers prematurely. Spores are formed in the black spots. These spores escape through the ruptured leaf epidermis when mature and are disseminated to other plants. The fungus survives from one season to the next on diseased leaves.

Control measures are not required but some observations on varietal resistance have been made. In Japan leaf smut appears to occur more frequently on early varieties. A similar pattern was observed in India. In USA differences in varietal susceptibility were found, but many of the practical varieties were susceptible.

Tilletia barclayana

Kernel smut

Distribution

Kernel smut occurs in Japan, USA, Burma, China, Fiji, India, Indonesia, Malaysia, Nepal, Pakistan, Philippines, Taiwan, Thailand, Vietnam, Guyana, Mexico, Trinidad, Venezuela and Sierra Leone (see CMI Distribution Map 75).

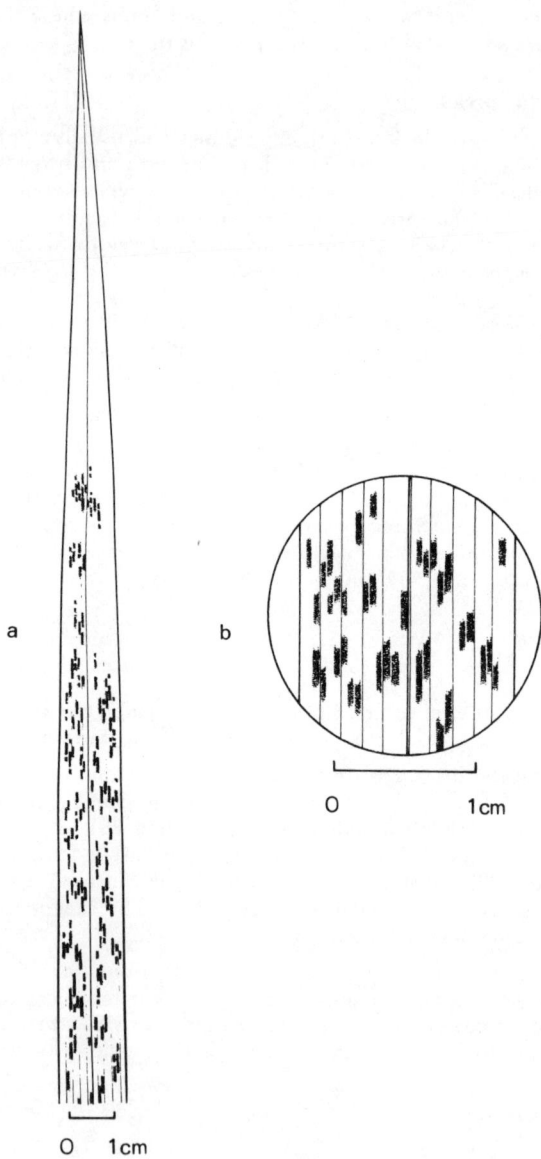

a

b

O   1cm

O   1cm

Fig. 25. Leaf smut, *Entyloma oryzae;* entire leaf showing
lesions (a) and lesions magnified showing minute black
slightly raised spots (b). (H.R.B.)

Taxonomy

The name now generally accepted for the causal agent of kernel smut is *Tilletia barclayana* (Basidiomycotina: Ustilaginales). The original name of the smut was *Tilletia horrida*. Later is was named *Neovossia horrida*. It has been suggested that the fungus causing rice smut is identical to *Neovossia barclayana*. The name *Neovossia horrida* is used by some authors.

Symptoms

This disease only affects the grains and is observed as small, black pustules and streaks which become more obvious as the grains ripen. Sometimes the whole grain is replaced by a powdery black mass of spores. These spores scatter onto other grains and leaves. Rain or heavy dew make the symptoms more obvious. It is characteristic of this smut, however, that only part of the kernel may be affected; in this case the kernel may be twisted to one side so that it protrudes between the glumes.

Development and spread

Chlamydospores survive for a year or more under normal conditions. Viable chlamydospores have been found after three years in stored grain. Chlamydospores germinate and produce many sporidia on the promycelium. The primary sporidia either produce secondary sporidia directly or produce a sporidium on which the secondary sporidia are borne. The secondary sporidia, which are sickle-shaped, are forcibly discharged into the air thus aiding dissemination and spread of infection.

It has generally been found that a high rate of nitrogenous fertilizer application, especially at a late stage of growth, increases incidence of smut. In USA additional factors favouring the disease are a high level of soil fertility and heavy clay soil. Weather conditions also affect incidence, frequent light rainfall and humid weather when the plants are flowering are conducive to attack.

Control

No positive evidence has been put forward to show that the disease is systemic. Unlike similar diseases, kernel smut is not seedborne and infection does not take place through the germinating seed. Seed treatment is therefore ineffective.

Kernel smut is not usually of economic importance and special control measures are not normally needed. Resistant varieties can be used where necessary. Little detailed information on varietal resistance is available, but in USA short and medium-grain varieties are in general more resistant than long-grain types. In India later-maturing varieties have been found to be more resistant than early or medium varieties.

Ustilaginoidea virens

False smut

Distribution

The pathogen is widespread in many rice-growing areas of Asia, Latin America and Africa but does not occur in Australia (CMI Distribution Map 347).

Taxonomy

The fungus causing this disease is *Ustilaginoidea virens* (Hyphomycetes). It was first described as a true smut and since then various names have been applied to the organism. The perfect state was discovered in 1934 and named *Claviceps virens*. More recently the name *Claviceps oryza-sativae* Hashioka (*nom. nov.*) has been suggested.

Symptoms

False smut is one of the most conspicuous diseases affecting rice but rarely causes significant losses. Generally the disease is more common and more serious in temperate and cooler mountainous rice-growing districts than in the hot tropical regions. The disease occurs more frequently in upland rice areas than in lowland areas when conditions are favourable. The disease is usually sporadically distributed in a field and only a few grains per ear are affected.

The symptoms of false smut are confined to the ears when individual grains develop into yellow or greenish velvety spore balls which become greatly enlarged at the later stages. The spore balls are covered by a membrane in the early stages, which bursts with further growth. These bodies are green on the surface and orange within. The glumes of the original grains may sometimes be seen on the surface of the false smut balls or they may be enclosed within the growth. High relative humidity and high rainfall favour the development of the disease which tends to be more serious on plants supplied with excess nutrients. A high incidence of false smut can be associated with bad management.

Control

Special control measures are not usually warranted. Spraying with fungicides during the period from booting to early heading has been shown to prevent false smut infection. In Japan most susceptible varieties are found amongst the late maturing varieties.

In India, 297 varieties were tested and 186 remained free from infection despite favourable environmental conditions.

Bibliography

ATKINS, J. G. *et al.* (1967). An international set of rice varieties for differentiating races of *Piricularia oryzae. Phytopathology* 57: 297–301.

AWODERU, V. A. (1974). Rice diseases in Nigeria. *PANS* 20 (4): 416–424.

BANDONG, J. M. and OU, S. H. (1966). The physiologic races of *Piricularia oryzae* Cav. in the Philippines. *Philippine Agriculturalist* 49 (8): 655—667.

BEDI, K. A. and GILL, H. S. (1960). Losses caused by the brown leaf-spot disease of rice in the Punjab. *Indian Phytopathology* 3(2): 161—164.

CAMPOS, G. L. (1971). La Piricularia (mal del cuello, fallada, hongo, etc). *Arroz* 38: 1—3.

CAROLIS, C. de (1971). Prime prove di resistenza in pieno campodi venticinque varietà italiane di riso all'agente patogeno del 'brusone' fogliare (*Piricularia oryzae* Cavara). *Il Riso* 20 (1): 171—175.

CHAKRABARTI, N. K. *et al.* (1966). The present position of physiologic races of *Piricularia oryzae* Cav. in India. *Bulletin of the Indian Phytopathological Society* 3: 102—109.

CHAUHAN, L. S. and VERMA, S. C. (1964). Bunt resistance of paddy varieties in Uttar Pradesh. *Science and Culture* 30: 201.

CHIU, R—J., CHIEN, C. C. and LIN, S. Y. (1963). Physiologic races of *Piricularia oryzae* in Taiwan. In *The rice blast disease*. International Rice Research Institute. pp. 245—255. Johns Hopkins Press, Baltimore, 1965.

CHOWDHURY, S. (1951). Studies in the bunt of rice (*Oryza sativa* L.). *Indian Phytopathology* 4: 25—37.

DEIGHTON, F. C. (1967). *Sphaerulina oryzina*, the perfect state of *Cercospora oryzae*. *Transactions of the British Mycological Society* 50: 499.

GALVEZ, E. G. E. and LOZANO, J. C. (1968). Identification of races of *Piricularia oryzae* in Colombia. *Phytopathology* 58: 294—296.

GHOSE, R. L. M., GHATGE, M. B. and SUBRAHMANYAN, V. (1960). *Rice in India.* Revised Edition. pp. 474. Indian Council of Agricultural Research, New Delhi.

GOTO, K. *et al.* (1967). US — Japan co-operative research on the international pathogenic races of the rice blast fungus, *Pyricularia oryzae* Cav., and their international differentials. *Annals of the Phytopathological Society of Japan* 33 (Extra Issue): 1—87.

HASHIOKA, Y. (1969). Rice diseases of the world — IV. Diseases due to Phycomycetes (Fungal Diseases — 1). *Il Riso* 17 (1): 279—290.

HASHIOKA, Y. (1969). Rice diseases of the world — V. Smuts (Fungal diseases — 2). *Il Riso*. 17 (1): 11—15.

HASHIOKA, Y. (1969). Rice diseases of the world — VI. Sheath spots due to sclerotial fungi (Fungal diseases — 3). *Il Riso* 17 (1): 111—128.

HASHIOKA, Y. (1969). Rice diseases of the world — VII. Diseases due to Sphaeriales, Ascomycetes (Fungal diseases — 4). *Il Riso* 17 (1): 309—338.

HASHIOKA, Y. (1971). Rice diseases of the world — VIII. Diseases due to Hypocreales, Ascomycetes (Fungal diseases — 5). *Il Riso* 20(1): 235—258.

HUANG, Y—T. and YU, C—M. (1973). Studies on seedling diseases of rice in the non-submerged seedling bed. 1. Control of seedling blight of rice by seed treatments. *Plant Protection Bulletin, Taiwan* 15 (4): 139—146.

KATO, H. (1974). Epidemiological aspect of sporulation by blast fungus on rice plants. *Japan Agricultural Research Quarterly* 8(1): 19—22.

KOZAKA, T. (1970). Pellicularia sheath blight of rice plants and its control. *Japan Agricultural Research Quarterly* 5(1): 12—16.

KRAUSE, R. A. and WEBSTER, R. K. (1972). The morphology, taxonomy and sexuality of the rice stem rot fungus, *Magnaporthe salvinii (Leptosphaeria salvinii).* *Mycologia* 64: 103—114.

LEE, L. F. and CHIU, K. Y. (1973). Control of damping-off disease of rice with fungicides mixed in seedlings soils. *Plant Protection Bulletin, Taiwan* 15 (2).

LEE, S. C. and MATSUMOTO, S. (1966). Studies on the physiologic races of rice blast fungus in Korea during the period of 1962–1963. *Annals of the Phytopathological Society of Japan* 32: 40–45.

MANIBHUSHANRAO, K. and DAY, P. R. (1972). Low night temperature and blast disease development on rice. *Phytopathology* 62: 1005–1007.

MATHUR, S. C. (1973). Studies in seedling blight of rice caused by *Sclerotium rolfsii*. *Il Riso* 22(3): 231–237.

OHATA, K. (1972). Ear blighting of rice plants and its control. *Japan Agricultural Research Quarterly* 6(3): 142–146.

OU, S. H. (1972). *Rice Diseases.* pp. 368. Commonwealth Agricultural Bureaux, Farnham Royal, UK.

OU, S. H. and AYAD, M. R. (1968). Pathogenic races of *Pyricularia oryzae* originating from single lesions and monoconidial cultures. *Phytopathology* 58: 179–182.

PADMANABHAN, S. Y., GANGULY, D. and CHANDWANI, G. H. (1966). Helminthosporium disease of rice. VIII. Breeding resistant varieties of early duration from genetic stock. *Indian Phytopathology* 19: 72–75.

PADWICK, G. W. (1950). *Manual of rice diseases.* pp. 198. Commonwealth Mycological Institute, Kew, UK.

PARK, C. S. and CHO, Y. S. (1972). Control of some seed borne organisms on rice with Dithane M-45. *Korean Journal of Plant Protection* 11(2): 109–111.

RAO, K. M. (1964). Environmental conditions and false smut incidence in rice. *Indian Phytopathology* 17: 110–114.

SINGH, R. A. and PAVGI, M. S. (1966). Varietal reaction to leaf smut of rice. *Indian Phytopathology* 19: 378–382.

SRIVASTAVA, M. P. and AHUJA, S. C. (1973). Stem rot resistance in rice varieties. *Il Riso* 22(3): 239–242.

SRIVASTAVA, M. P. and MAHESHWARI, S. K. (1971). Varietal resistance to brown leaf spot. *Il Riso* 20(1): 271–273.

SRIVASTAVA, M. P., MAHESHWARI, S. K. and KANG, M. S. (1974). Multiple resistance for diseases in rice varieties. *Il Riso* 23(3): 283–285.

TEMPLETON, G. E. (1964). Kernel smut of rice as affected by nitrogen. *Arkansas Farming Research* 12:12.

TEMPLETON, G. E., JOHNSTON, T. H. and HENRY, S. E. (1960). Kernel smut of rice. *Arkansas Farming Research* 9:10.

TEMPLETON, G. E. and WORAWISITTHUMRONG, A. (1963). Leaf smut of rice. *Arkansas Farming Research* 12:4.

TULLIS, E. C. and JOHNSON, A. G. (1952). Synonymy of *Tilletia horrida* and *Neovossia barclayana*. *Mycologia* 44: 773–788.

UMEDA, Y. (1973). Hinosan, a fungicide for control of rice blast. *Japan Pesticide Information* 17: 25–28.

WEBSTER, R. K., HALL, D. H., BOLSTAD, J., WICK, C. M., BRANDON, D. M., BASKETT, R. and WILLIAMS, J. M. (1973). Chemical seed treatment for the control of seedling disease of water-sown rice. *Hilgardia* 41(21): 689–698.

YAMAGUCHI, T. (1971). 1970. Evaluation of candidate pesticides (B–I) Fungicides: Preparations for controlling rice crop diseases. *Japan Pesticide Information* 8: 17–20.

YAMAGUCHI, T. (1972). 1971 evaluation of candidate pesticides (B–I) Fungicides: Rice. *Japan Pesticide Information* 12: 15–17.

YAMAGUCHI, T. (1973). 1972 evaluation of candidate pesticides (B—I) Fungicides: Rice. *Japan Pesticide Information* 16: 17—21.

YAMAGUCHI, T. (1974). Development of rice blast control techniques:— Japan. *Japan Pesticide Information* 18: 5—9.

YAMAGUCHI, T. (1974). Control of rice diseases by fungicides applied to submerged water in Japan. *Japan Agricultural Research Quarterly* 8(1): 32—36.

BACTERIAL DISEASES

Xanthomonas oryzae
Bacterial leaf blight

Distribution

Bacterial blight has been known for a long time in Japan and is also widespread in other Asian countries. The recognition of its importance in tropical countries such as India, the Philippines and Indonesia is only comparatively recent. Bacterial leaf blight has also been reported from Australia. However, it is thought that the Australian isolate is probably a different strain of the bacterium and has been indigenous in Australia for some time.

Symptoms

First symptoms are usually noticed at about heading stage but in severe cases they may appear earlier. Symptoms usually appear as water-soaked lesions in the form of stripes. The lesions have a wavy margin and turn yellow after a few days. They usually start at the leaf margins (Fig. 16) but may start anywhere on the leaf blade at the site of an injury. As the disease progresses the lesions spread over the whole leaf blade turning it white (Plate 6). Lesions also occur on the leaf sheath in susceptible varieties. Symptom expression on the leaf blades varies with varietal susceptibility. In more resistant varieties the lesions remain yellow with only slow development of necrosis whereas in susceptible varieties the development of necrosis is very rapid and the affected leaves roll up as they become necrotic. Young lesions produce milky drops of bacterial exudate which are most commonly observed in the early mornings. Droplets can be detected by drawing the leaves through the fingers and feeling the stickiness. These drops dry up and easily fall off into the paddy water.

In the tropics bacterial blight can result in a systemic infection. This expression of the disease is referred to as 'kresek'. 'Kresek' symptoms are usually observed one or two weeks after transplanting. Infection takes place through wounds at transplanting and moves to the growing point. In older plants after the growing point becomes infected the youngest leaf has a pale yellow colour. When the practice of cutting off the leaf tips at transplanting has been carried out infection often takes place through the resulting wounds and spreads downwards through the leaves which necrose as the infection progresses. Death often results when young plants are infected. 'Kresek' symptoms often superficially resemble damage caused by stem borers. Systemic infection by bacterial blight can be diagnosed by the yellowish-white bacterial fluid which oozes from the vascular bundles when an infected stem is cut and squeezed. Bacterial ooze is also produced when infected leaves are cut and immersed in water.

Losses are usually only serious where high fertilizer rates are used. Up to 400,000 ha in Japan are affected by bacterial blight annually. In severely affected fields losses are between 20 and 30%. More severe losses have been caused in Indonesia and the Philippines and bacterial blight is of major importance in India. Damage tends to be worse in the tropics.

Development and spread

Bacterial blight is principally a vascular disease and, as discussed above, often enters through wounds caused by transplanting. Infection can also take place through the hydathodes and often occurs through wounds resulting from storm damage. Thus, after a typhoon infection from a few plants may spread to an entire field. The drops of bacterial exudate produced on the leaves can quickly pass the bacteria to other plants direct or via the irrigation water. Irrigation water can carry the disease from one field to another. Evaluation of the number of bacteriophages in the irrigation water can be used to estimate the bacterial population but is not always strictly accurate.

The bacteria can overwinter in soil and seed but seed transmission is not considered to be important. Weed hosts (e.g. *Leersia sayanuka, L. oryzoides* and *Zizania latifolia*) play an important part in the desease cycle as do diseased straw and stubble.

Bacterial blight is prevalent in badly drained swampy areas where flooding is likely.

Control

Cultural methods of minimising the risk of attack by bacterial blight are avoidance of deep flooding; destruction of crop remains by burning or thorough ploughing; and elimination of weed hosts. Careful application of fertilizer to avoid excess nitrogen can also lessen the likelihood of a disease outbreak. Nitrogenous fertilizer can be applied by small split applications.

Numerous resistant varieties have been developed but owing to the diversity of bacterial strains varieties resistant in one area are not necessarily resistant in another. Resistance to 'kresek' is not correlated to resistance to other symptoms. Approximately 7,000 varieties have been screened for resistance to bacterial blight at IRRI. The most resistant are Zenith, TKM 6, Malagkit Sungsong, Wase Aikoko, Mortgage Lifter, Early Prolific, Bluebonnet X Rexark, Lacross X Zenith and Sigadis. Local advice should be sought to discover the best variety for a particular area.

Chemical control is often not economic especially as the disease is well controlled by the use of resistant varieties.

Antibiotics for control of bacterial leaf blight are cellocidin and chloramphenicol. Synthetics used for control are dithianon, nickel dimethyldithiocarbamate, phenazine oxide and fentiazon.

Phenazine oxide or fentiazon should be applied after typhoon damage.

Phenazine oxide (10% a.i.) is recommended at a rate of 100 g/50—80 l of water applied at up to 1,000—1,200 l/ha. Applications should be started at the first sign of infestation and repeated, according to conditions at approximately one week intervals.

Nickel dimethyldithiocarbamate (65% a.i.) is applied at 225 g in 100—130 l of water applied at 1,000—1,300 l/ha. The 8% a.i. dust formulation should be applied at 30—40 kg/ha.

Applications to the nursery bed are very effective. Under severe conditions applications to the nursery bed and the field are needed. Applications to the nursery bed should be made two or three times at seven to ten day intervals with the last application just before transplanting. Field applications should be started as soon as there is a threat of the disease and repeated at seven to ten day intervals.

TF-128 and TF-130 (thiadiazole compounds) were shown to be highly effective against bacterial leaf blight but owing to doubt over the carcinogenicity of the residues practical use ceased.

Xanthomonas translucens f. sp. oryzicola

Bacterial leaf streak

Bacterial leaf streak is also an important bacterial disease but generally causes less damage than leaf blight. Bacterial leaf streak is widespread in tropical Asia having been reported in the Philippines, South China, Thailand, Malaysia, India, Vietnam, Indonesia, Cambodia and Australia. It is not known to occur in temperate areas.

Symptoms

Bacterial leaf streak first appears as fine, interveinal, long or short lines on the leaf blade, water-soaked and greyish (Fig. 16). Close examination reveals minute, yellowish beads of dried bacterial exudate on these lesions. The lesions extend and coalesce to form larger patches and become yellow. Eventually a large portion or often the whole of the leaf blade becomes yellow or dirty white (Plate 7). At this late stage the symptoms are difficult to distinguish from those of bacterial leaf blight.

The bacterial exudates formed on the surface of the leaf spread the disease from plant to plant and from one field to another in the same way as bacterial blight is spread.

Control

The use of resistant varieties offers the best means of control. Fungicidal control tends not to be economic as the bacterium attacks the mesophyll of the leaf in contrast to bacterial leaf blight which is a vascular disease.

Varieties showing resistance include BJ1, DD89, DD113, DNJ-142, DV29, DV52, DV98, DZ192, Hasikalme, Blue Rose, Milbuen 5 (3), Charnock, Zenith, Tetep, H4, S67 and CO4.

Indica varieties are much more susceptible than japonica varieties.

Bibliography

ALDRICK, S. J., BUDDENHAGEN, I. W. and REDDY, A. P. K. (1973). The occurrence of bacterial leaf blight in wild and cultivated rice in northern Australia. *Australian Journal of Agricultural Research* 24: 219–227.

BRADBURY, J. F. (1971). Nomenclature of the bacterial leaf streak pathogen. *International Journal of Systematic Bacteriology* 21:72.

HASHIOKA, Y. (1969). Rice diseases in the world – III. Bacterial diseases. *Il Riso* 18(1): 189–205.

OU, S. H. (1972). *Rice Diseases.* pp. 368. Commonwealth Agricultural Bureaux, Farnham Royal, UK.

PITKETHLEY, R. N. (1970). A preliminary list of plant diseases in the Northern Territory. *Technical Bulletin. Primary Industries Branch, N. T. Administration, Darwin* No. 2. pp. 30. (cyclostyled).

RAO, Y. P., SHEKHAWAT, G. S., MOHAN, S. K. and REDDY, P. R. (1972). Evaluation of rice varieties for resistance to bacterial leaf-streak. *Indian Journal of Agricultural Sciences* 42(6): 502–505.

VIRUS AND MYCOPLASMA DISEASES

This section deals with rice diseases caused by viruses, mycoplasmas or organisms which are thought to be mycoplasmas.

Viruses are submicroscopic particles which can multiply in living cells. The importance of insect vectors and the infectivity of virus diseases distinguishes them from disease caused by fungi and bacteria. Viruses are particulate whereas mycoplasmas are minimal reproductive units lacking a cell wall.

Rice viruses are transmitted mechanically, by insects and through soil. In some virus vector relationships transovarial passage occurs. This is when the virus is transmitted from one insect generation to the next through the egg. Transstadial passage is when the insect does not lose the virus even after moulting.

There are 12 diseases in this section and the following key will aid in their identification.

KEY FOR CLASSIFYING RICE VIRUS DISEASES*

A_1 Plants showing inconspicuous stunting, but reduced tillering

   B_1 Upright growth habit, premature death, orange-coloured and rolled leaves ....ORANGE LEAF

   B_2 Spreading growth habit, oval to oblong faint chlorotic patches or fine faint mottling on leaves, brown necrotic lesions on basal parts of culms at later stages ....NECROSIS MOSAIC

A_2 Plants showing stunting and reduced tillering

   C_1 Leaves with chlorotic spots and white stripes

      D_1 New leaves not unfolding properly but twisted and droopy ....STRIPE

      D_2 New leaves unfolding normally ....HOJA BLANCA

   C_2 Leaves with mottling and yellowish streaks, crinkling of the first newly formed leaves when infected at an early stage of growth..........................YELLOW MOTTLE

   C_3 Leaves with yellow or yellow-orange discolouration

      E_1 Virus particles are bullet-shaped and persist in the vector   TRANSITORY YELLOWING

      E_2 Virus particles are spherical or polyhedral and do not persist in the vector   TUNGRO (leaf yellowing, penyakit merah, and yellow-orange leaf)

      E_3 Mycoplasma probably the causal organism, transmission unknown ....GIALLUME

A_3 Plants showing severe stunting and excessive tillering

   F_1 Galls on leaves and culms   BLACK-STREAKED DWARF

   F_2 No galls

      G_1 Leaves with chlorotic to whitish specks forming interrupted streaks.................................. DWARF

      G_2 Narrow, stiff, light-green leaves often with rusty spots.............................. GRASSY STUNT

      G_3 Leaves showing general chlorosis ....YELLOW DWARF (padi jantan)

* Reproduced from Ling (1973)

Black-streaked dwarf

Distribution

Black-streaked dwarf is only known to occur in Japan and usually yield losses are negligible.

Symptoms

Diseased plants are stunted and the leaves are darker green than normal. The most typical feature is the appearance of galls on the leaves; these are elongated swellings along the major veins and occur on the lower side of the leaf blade (Fig. 26). They are also found on leaf sheaths and culms. The galls may break through the epidermis, when the dark streaks from which the name of the disease is derived are formed. Leaf blades may be twisted. The disease is usually not lethal but only a few small panicles or none at all are produced. The panicle stalks are shorter than normal so that the panicles do not emerge fully. Any grains produced are often discoloured with dark brown spots.

The degree of growth retardation is determined by the age of the plant when infection takes place. A plant inoculated at the three leaf stage suffers a height reduction of 20%, whereas inoculation at the 14 leaf stage does not result in height reduction.

The virus has a wide host range. It affects maize, wheat and barley in the field and can cause serious damage to maize. Numerous graminaceous weeds are hosts. The water fox-tail, *Alopecurus aequalis* var *amurensis* is an important alternative host.

Black-streaked dwarf is transmitted by *Laodelphax striatellus, Unkanodes sapporonus* and *U. albifascia. L. striatellus* (the small brown planthopper) is the main vector. *U. sapporonus* usually lives on maize, wheat and barley and is therefore not important in the disease cycle on rice.

Control

Control of the disease depends on eliminating *L. striatellus*. This is difficult as the planthoppers continually migrate into the rice fields from their overwintering places in winter crops (e.g. wheat and barley) or wild grasses. The black-streaked dwarf virus does not make transovarial passage so the first generation nymph that appears is nonviruliferous. However the nymphs soon acquire the virus from winter crops.

By planting wheat and barley towards the end of the optimum period the chances of transmitting the virus from rice to the winter cereals is decreased. For information on vector control the reader should refer to the section on stripe.

Several varieties of rice are resistant to black-streaked dwarf. The resistance of the variety Tetep is controlled by a major dominant gene. This resistance is being used in programmes for breeding resistant varieties.

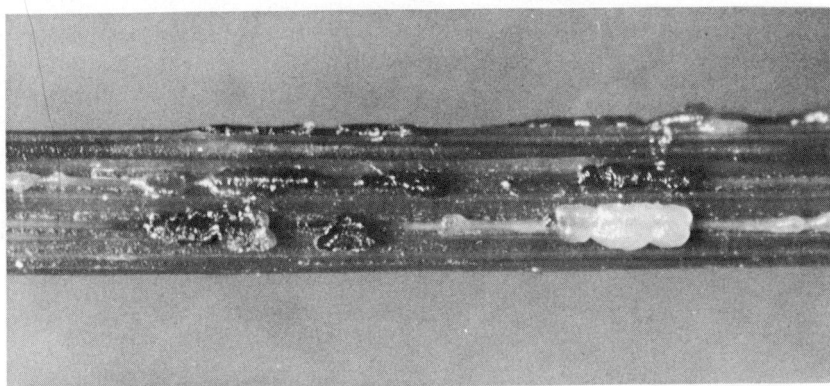

Fig. 26. Portion of rice leaf showing galls caused by black-streaked dwarf virus disease. (Dr A Shinkai)

Dwarf

Distribution

Dwarf (sometimes referred to as rice stunt) appears to be limited to Japan and Korea. It was the first virus disease of rice to be identified and the first plant virus disease found to be insect transmitted.

Symptoms

The above-ground parts of affected plants are markedly stunted and root growth is less than in healthy plants. There is an excessive production of tillers, which are very small (Fig. 27). On the leaves can be seen conspicuous yellow or white specks along the veins; these specks coalesce to form interrupted streaks. Specks and streaks are more abundant on later leaves and may also occur on leaf sheaths.

Plants affected by dwarf are not usually killed, but remain green after healthy plants have ripened. Panicles may be completely suppressed and if they are formed they are small and yield only a few small, poorly filled grains, often bearing dark spots.

Transmission

Dwarf is transmitted by the leafhoppers: *Nephotettix cincticeps, N. nigropictus* and *Recilia dorsalis. N. cincticeps* is the most important vector. Rice dwarf can make transovarial passage but the rate is not as high as for stripe. Most insects retain their infectivity for life. Damage and infection caused by the overwintered adult and 1st nymphal generation are usually slight. The 1st generation adult transmits the virus actively. The 2nd generation nymph causes great damage in early cultivation.

Fig. 27. Symptoms of dwarf virus disease plant on right showing severe stunting and retardation compared with healthy plant of the same age on left. (Dr A Shinkai)

Host range

Fifteen graminaceous plants, including wheat and barley, have been found susceptible to this dwarf virus, and in the field *Echinochloa crusgalli* and *Paspalum thunbergii* frequently show symptoms and can act as reservoirs of infection. Although wheat and barley have been infected by inoculation, natural infection of these hosts has not been found. Plants other than rice are not considered important as reservoir hosts.

The disease has become serious in Japan since earlier planting became popular, because the vectors multiply and can transmit the virus more readily in early planted rice.

Control

Chemical control is thought to be necessary when over 15% of the hills are infected. Where dwarf and yellow dwarf occur together insecticides should be applied by air in autumn or early spring (Japan) to reduce the insect population before transplanting. Suitable insecticides include carbaryl, BPMC, fenitrothion, MTMC and propoxur.

When necessary, chemical control after transplanting should be applied as early as possible. Control is similar to that for stripe but applications can be made over a smaller area as the vectors are not so widely dispersed. Insecticides such as vamidothion, malathion, BPMC, and carbaryl can be used.

Varieties resistant to rice dwarf are available.

Giallume

Giallume is an Italian expression for yellowing. Diseased plants are stunted with reduced tillering and yellow leaves. Symptoms are similar to those of yellow dwarf except there is yellowing rather than general chlorosis and tillering is slightly reduced rather than increased. The disease is caused by a virus which is very similar to barley yellow dwarf virus. One report assumes that giallume and rice yellow dwarf have the same causal agent. Giallume is transmitted by the aphid *Rhopalosiphum padi*. There is transstadial passage. *R. padi* can transmit the virus to rice, barley, oats, wheat, maize and cut grass (*Leersia oryzoides*). During the winter the virus can survive in the vector and alternative graminaceous hosts. Possible control methods are insecticide control of the vector, control of graminaceous weeds which may harbour the virus and use of resistant varieties.

Grassy stunt

Distribution

Grassy stunt was first seen in 1963 in the Philippines and has been most studied there, but it is also reported from Sri Lanka, Thailand, India and Indonesia. The disease commonly follows an outbreak of brown planthopper. Since its discovery it has been found in several areas in the Philippines and other countries. More than 50% infection has been found in some fields. Inoculation experiments indicate that grassy stunt is potentially serious. After early inoculation practically no grain can be harvested. Grassy stunt destroyed the 1974—75 rice crop on at least 200,000 ha of Indonesia's rice lands.

Symptoms

The disease is characterised by severe stunting. In the variety IR8 a 55% reduction in height results when the plants are inoculated 15 days after germination. The reduction in plant height decreases with later inoculation. Numerous small tillers are produced, giving the typical grassy appearance (Fig. 28). Plants have an erect growth habit, and leaves are pale green or yellow and bear numerous rusty-coloured spots. On older leaves these spots, spreading, give a bronze colour to the plants. Diseased plants are not usually killed but no panicles are formed, or those which are produced are small and sterile. Symptoms in general bear considerable resemblance to those of yellow dwarf but the diseases are distinct.

Previously it was thought that the disease was caused by a virus but it is now suspected that the disease may be caused by mycoplasma-like organisms (Plate 8).

Transmission

Grassy stunt is transmitted by brown planthopper, *Nilaparvata lugens*. About 20—40% of the population are active transmitters of the causal agent. The causal agent is not transmitted by seed or transovarially but transstadial passage occurs. Fifteen rice species are hosts of the causal agent.

Fig. 28. Grassy stunt virus disease, infected plant. Note very severe stunting and extra tillering. (H.R.B.)

O   5cm

Control

Pot experiments on the use of tetracyclines for control of the disease have so far been unsuccessful.

Several varieties have been found to be tolerant to the disease but IR-26 is resistant. A strain of wild rice, *Oryza nivara* is highly resistant to the disease but is susceptible to the vector. Some varieties, e.g. Mudgo are resistant to the vector but susceptible to the disease. It is hoped that the resistant characteristics can be combined by plant breeding.

Hoja blanca

Distribution

Hoja blanca means white leaf in Spanish. This disease is known only in the Western Hemisphere, occuring in North, Central and South America (CMI Distribution Map 359). Yield losses vary between slight and nearly complete. In 1956 there was a 25% yield loss in Cuba and over 50% loss in Venezuela.

Symptoms

Affected plants are much reduced in size. Elongated white stripes appear on the leaves, or when more severely attacked the leaves may be almost entirely white. Sometimes they show a diffuse mottling, which may also appear on the culms.

Panicles are considerably shorter than normal and often do not emerge completely from the boot leaf. Few or no grains are produced, and the panicle, being light remains upright. If plants are infected at an early stage of growth they may be killed by the disease. Symptoms may vary according to the rice variety. The symptoms are more severe when the plants are young at the time of infection.

Transmission

Hoja blanca is transmitted by *Sogatodes orizicola* and *Sogatodes cubanus*. The shortest acquisition feeding time is 15 min. Only about 5–15% of a population of leafhoppers are normally able to transmit the disease; and the insects are somewhat inactive, which helps to some extent to limit the rate of spread of infection. There is some evidence that the virus passes transovarially to the progeny of infected females.

Several grasses have been shown to be susceptible to the virus, the most important being *Echinochloa colonum*. This species is a common weed in and around rice fields, and is able to grow throughout the year in warmer areas. It thus acts as a permanent reservoir of infection for the reinfection of the rice crops. *S. cubanus* feeds mainly on *E. colonum* and does not favour rice but enough feeding on rice appears to occur to allow transmission of the virus from the grass to rice, although conflicting reports on this point have been published. *S. orizicola* feeds by preference on rice and is therefore an important means of spread of hoja blanca within the crop.

No transmission by seed, soil, or mechanical inoculation has been found. The main factor affecting the severity of hoja blanca is the population of leafhoppers in and around rice fields. This population varies greatly at different times of the year, which in turn affects the rate of spread of the virus in the crop.

Control

Varieties resistant to hoja blanca are available. Resistance to the disease is not necessarily associated with resistance to *S. orizicola*. Chemical control can be partially successful when used in conjunction with resistant varieties.

Necrosis mosaic

Necrosis mosaic, which was previously called rice dwarf, is a soil-borne virus disease reported from Japan. The main symptoms are spreading growth habit, mosaic of leaf blades (especially lower ones) and necrotic lesions on basal portions of culms and leaf sheaths. The disease usually occurs in upland seedbeds. Up to 47% loss of grain yield has been reported. Infected plants are more susceptible to blast. The disease is transmitted mechanically and by soil. As a control measure infected seed beds should not be reused. Where this is not possible the soil can be disinfected using chloropicrin, methyl bromide or quintozene. The varieties Omachi and Shinonome-mochi are resistant to the disease.

Orange leaf

Distribution

Orange leaf was first noticed as a distinct disease in Thailand in 1960. It was later recorded in the Philippines and Sri Lanka and was identified as a new virus disease of rice in 1963. In the field it is, however, usually scattered and appears to cause comparatively little damage. Orange leaf often appears in the same field as tungro and the symptoms are easily confused.

Symptoms

Affected plants show discolouration of the leaves. At first discolouration appears as stripes but when the plants are about one month old the leaves have an overall golden-yellow to deep orange colour. Infected leaves roll up and die (Fig. 29, Plate 7). Diseased plants have fewer tillers than normal and root development is poor. Growth is arrested and death often ensues. If the plant survives to produce panicles these do not emerge properly and are sterile. Seedlings inoculated at the 3-leaf stage die within 3–4 weeks.

The disease is caused by a virus and the leafhopper *Recilia dorsalis* is the only vector. The virus is not transmitted by soil or seed or mechanically but appears to be persistent in the vector. Seven to 14% of the population are active transmitters of the virus. Minimum acquisition period is 5 h and the insects remain infective throughout their lives. No alternative hosts of the virus have been discovered. Some resistant and tolerant rice varieties are available.

Rice yellow mottle virus (RYMV)

This disease was first reported from Kenya around Lake Victoria in 1970 and appears to be confined to this area. Symptoms are stunting and reduced tillering of the plant. Leaves become crinkled, mottled and have yellow streaks. Panicles are malformed and only partially emerge. In severe cases the plant may die. RYMV is mechanically transmissible to rice and *Oryza barthii* and *O. punctata*. Tests for other gramineaceous hosts have so far proved negative. The virus has been found in guttation fluid and in the irrigation water of heavily infected fields. The beetle *Sesselia pusilla* was first recognised as a vector. Later the beetles *Chaetocnema pulla*, *Trichispa sericea* and *Dicladispa (Chrysispa) viridicyanea* were found to be vectors.

Stripe

Distribution

Stripe disease is widely distributed in Japan and South Korea (CMI Distribution Map 359). Since the introduction of earlier planting in the early 50s the disease has increased in importance. It is often found together with black-streaked dwarf. Yield reduction decreases with plant age at infection.

Fig. 29. Orange leaf virus disease, infected plant. Note infected leaves rolling at the tips and reduced tillering. (H.R.B.)

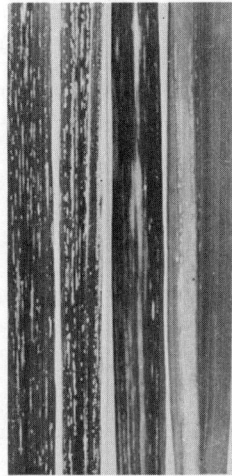

Fig. 30. Symptoms of stripe virus disease on rice leaves. Two leaves on left from upland rice, two on right from lowland transplanted rice. Note small chlorotic lesions on leaf blades merging into 'stripes'. (Dr A Shinkai)

Symptoms

A characteristic feature of stripe is the failure of emerging leaves to unfold properly; they become twisted, droop, and are elongated. They are also chlorotic, either generally or in stripes with diffuse margins. The chlorotic areas later become necrotic and the leaves die. Leaf sheaths may also show a chlorotic mottling (Fig. 30). Plants are more susceptible at the seedling or early tillering stage and if infection is at this early stage the whole plant often dies. If it does not, it is stunted and the number of tillers is much reduced, effects which become less the later infection takes place.

Affected plants, even if they survive, produce few if any panicles and these have malformed spikelets and often emerge incompletely from the boot leaf. In slight infections few symptoms may be seen except that many of the grains may be misshapen and only partially filled.

Transmission

Stripe disease is transmitted by *Laodelphax striatellus*, *Unkanodes sapporonus* and *U. albifascia*. Transovarial passage has been shown to occur in all three vector species.

The main factor affecting the prevalence and severity of stripe is the population density of the first generation leafhoppers and their time and mode of occurence, disease incidence being highest when large numbers of the insects invade the rice fields soon after transplanting.

Host range

Numerous graminaceous species, including wheat, barley and maize, can act as hosts to the virus. Symptoms on these alternative hosts tend to be mild.

Control

Some reduction of stripe virus infection of rice is possible by using insecticides to control the vectors. However, such control is difficult because of the continual migration of the vectors into the rice fields from their overwintering places. Control methods are not effective if used only over small areas. In Japan aerial spraying during March and April can be used to reduce the population of over-wintering adults. U.l.v. or fine granule formulations are recommended in order to reduce effects on mammals. If there is a large population of 1st generation nymphs during the period May—June and early control has not been performed then extensive control at this stage is recommended. BPMC, propoxur, fenitro-thion, MTMC and carbaryl are suitable for control at these two stages.

Control of the 1st generation adult in the paddy field should be carried out over as wide an area as possible. Numerous insecticides are available for control at this stage. For control of the 2nd generation nymphs which are less mobile effective results can be obtained over small areas using submerged application of granules. Care must be taken to follow the recommendations on application near harvesting and heading.

Numerous varieties resistant to stripe are available. Where stripe and black-streaked dwarf occur together it is best to use stripe resistant varieties. The variety Mineyutako is highly resistant to stripe. The genetics of resistance varies in the different resistant varieties.

Transitory yellowing

Distribution

Transitory yellowing is only known to occur in Taiwan. Although the disease may have existed in Taiwan for a long time it did not attract attention until there was an outbreak in 1960 in the south of the island. The disease is now widespread in Taiwan. Yield losses are usually 25—32%. Panicles are smaller and fewer and there are more unfilled grains than in healthy plants.

Symptoms

In some aspects symptoms of tungro and transitory yellowing are similar. Leaves become discoloured from the tip, lower leaves showing more intense discolour-ation; the colour is yellow or orange buff. Rusty brown specks may also appear on the leaves but are of doubtful diagnostic value. The number of tillers is reduced and plants are slightly stunted. Sometimes later-produced leaves may show no symptoms, and as the yellowed leaves dry up and die all signs of the disease may disappear, giving the appearance of complete recovery, as suggested by the name given to the disease. Symptoms differ with variety. Large round inclusion bodies can be observed within parenchyma cells and around xylem vessels etc.

Transmission

Transitory yellowing is transmitted by *Nephotettix nigropictus*, *N. cincticeps* and *N. virescens*. There is no evidence of transmission by seed, soil or mechanical means. Percentage of active transmitters is 41–65% for *N. nigropictus*; 35–71% for *N. cincticeps* and 47% for *N. virescens*. There is transstadial passage but no transovarial passage. No alternative plant hosts of the virus have been found.

Control

The planting of resistant varieties appear to offer the best hope of controlling transitory yellowing. Varieties Chu-tze, Chung-lin-chun, Wu-ku-chin-yu, Hu-lu-tuen and Kao-hsiung 22 are highly resistant.

Tungro

Distribution

This disease is also called peny.akit merah, yellow-orange leaf, leaf yellowing and mentek. The names vary in different parts of the world but it now appears that they all refer to the same disease.

Tungro occurs in Philippines, Malaysia, Thailand, India, Indonesia and Bangladesh. A disease with similar symptoms to tungro has appeared in Japan causing yield losses up to 4%. If the disease is tungro then differences in symptoms may be due to the different rice varieties grown in Japan. Yield loss is more serious when the plants are young. In inoculation experiments the yield of plants inoculated when 15 days old was reduced by 68%. Tungro has caused serious crop losses in Sumatra. Plants infected with tungro take longer to flower. For example, flowering is delayed by four weeks in IR8 infected at 15 days (Plate 8).

Symptoms

Affected plants are stunted, but the number of tillers produced is normal (Fig. 31). Leaves are markedly discoloured, the discolouration varying considerably with the variety and age of the plant. Usually the colour of affected plants is yellow or orange, the change in colour spreading downwards from the tips of the leaves, sometimes for a short distance but frequently to involve the whole leaf blade. Small rust-coloured spots may appear on older leaves. The progress of symptoms depends on the variety. In some the leaf symptoms persist throughout the life of the rice plant and the plant remains severely stunted, in others all symptoms disappear after a time, giving the appearance of recovery from the disease.

Flowering is delayed in infected plants and production of grain is affected. Diseased plants have fewer spikelets than healthy plants, a higher proportion of

Fig. 31. Tungro virus disease, infected
plant. Note stunting of plant and
drooping of leaves. (H.R.B.)

O   5cm

Fig. 32. Symptoms of yellow dwarf
virus disease, diseased plant on left
showing stunting and chlorosis of the
entire plant compared with healthy
plant of the same age on right. (Dr A
Shinkai)

empty grains, lower grain weight, and fewer grains which may be discoloured,
and the result may be a considerable reduction in yield.

In resistant varieties leaf discolouration is only partial and leaves produced later
may not show discolouration, however the plants are usually slightly stunted.
In varieties with moderate resistance discolouration often gradually disappears at
later stages and the plants yield well.

Plants infected at late stages of growth may not show symptoms before harvest,
however the symptoms may appear in the ratoon crop.

Transmission

There is no evidence of seed or soil transmission of tungro. Vectors of the virus
are *Nephotettix virescens, Recilia dorsalis, N. parvus* and *N. malayanus*. A
small percentage of *N. nigropictus* is able to transmit the virus but the most
important vector is *N. virescens*.

The virus persists for only a short time in the vector. There is no transovarial
passage and infective nymphs lose their infectivity after moulting.

Host range

Several species of grass and wild rice are hosts of tungro virus.

Control

Resistance to tungro appears to be associated with resistance to the vector. A wide selection of varieties resistant to tungro and/or its vector are available.

For control it is best to combine insecticidal control of the vector in seedling rice with the use of resistant varieties. Further information on chemical control can be found in the section on dwarf and in the insect section of the manual.

Yellow dwarf

Distribution

Yellow dwarf was first reported from Japan in 1919. The disease is widely distributed in Asia. Field distribution is sporadic and losses are not usually high. However ratoon growth and second crops can become heavily infected and yield losses of up to 50% have been reported. Yield loss of individual plants is determined by plant age at the time of infection. Overall annual losses of up to 10,000 t of grain have been estimated for Japan, with about 20,000 ha affected.

Yellow dwarf is caused by a mycoplasma-like organism, which may also be the causal agent of giallume.

Symptoms

As in dwarf, there is marked stunting and tillering is excessive; but leaf symptoms are distinct, affected leaves having an overall pale green or pale yellow colour (Fig. 32, Plate 8). If plants are infected early they usually die before maturity, and even if they do survive no panicles are produced or only a small number, with no grains. With later infection there may be little sign that the plant is diseased, but regrowth from stubble will show typical chlorosis.

Transmission

Yellow dwarf has been shown to be transmitted by seed.

Insect vectors are *Nephotettix nigropictus, N. virescens, N. cincticeps, N. parvus* and *N. malayanus.* The proportion of individuals in a population capable of transmitting the virus is much higher in yellow dwarf than in dwarf, and they can continue to transmit the disease throughout their life; but the disease is not transmitted transovarially, as is the case with dwarf.

Alopecurus aequalis, Glyceria acutiflora and *Oryza cubensis* have been said to be hosts of the disease but are not important in its perpetuation.

Control

Chemical control is needed when the percentage of diseased hills in the previous year's ratoon is more than 10%. Chemical control methods are the same as those for dwarf.

Varieties with resistance to yellow dwarf are available. Resistance is controlled by a dominant or incompletely dominant major gene.

Bibliography

AMICI, A., BALLESTEROS, R., BATALLA, J. A. and BELLI, G. (1970). Corpi riferibili a micoplasmi in piante di riso spagnolo offetto da 'enrochat'. *Rivista di Patologia Vegetale* 6: 247–253.

BAKKER, W. (1970). Rice yellow mottle, a mechanically transmissible virus disease of rice in Kenya. *Netherlands Journal of Plant Pathology* 76: 53–63.

BAKKER, W. (1971). Three new beetle vectors of rice yellow mottle virus in Kenya. *Netherlands Journal of Plant Pathology* 77 (6): 201–206.

BELLI, G., CORBETTA, G. and OSLER, R. (1975). Ricerche e osservazioni sull'epidemiologia e sulle possibilità di prevenzione del 'giallume' del riso. *Il Riso* 24 (4): 359–363.

ISHII, M. (1973). Control of rice virus diseases. *Japan Pesticide Information* 17: 11–16.

LING, K. C. (1972). *Rice virus diseases.* pp. 134. International Rice Research Institute, Los Banos, Philippines.

LING, K. C. (1973). *Synonymies of insect vectors of rice viruses.* pp. 29. International Rice Research Institute, Los Banos, Philippines.

OU, S. H. (1972). *Rice Diseases.* pp. 368. Commonwealth Agricultural Bureaux, Farnham Royal, UK.

OVERWATER, C. (1960). Tien jaren Prins Bernhard polder, 1950–1960. *De Surinaamse Landbouw* 8: 159–218.

POH, T. W. and ANG, O. C. (1974). Studies of penyakit merah disease of rice IV. Additional hosts of the virus and its vector. *The Malaysian Agricultural Journal* 49 (3): 269–174.

PHYSIOLOGICAL DISEASES

It is not within the scope of this Manual to deal with the complex issues involved in the nutritional requirements of rice. Fertilizer usage will not be discussed as a separate topic, but in some cases rice may show symptoms which resemble those of a fungus or virus disease because of a physiological disorder due to a nutritional deficiency. The major physiological diseases and their symptoms are briefly described below; along with the soil conditions which are responsible for their occurence.

It should be remembered that the aims of both pest control and fertilizer usage are identical, i.e. to achieve the maximum possible rice yield. If insects, diseases and weeds are controlled the greatest benefit will be obtained from any fertilizers applied to the crop. If fertilizers are used alone with no attempt at pest control their potential cannot be fully realised as the yield will be reduced by pest losses.

Fertilizer application must always be made with due regard to the local conditions of soil, climate and variety and the amount used is also linked to the size of yield required.

A fairly safe general rule for all fertilizer use is that it is only worth while using fertilizers if the return is economic. With a poor variety grown for a local market, a long distance from a fertilizer factory application is not worthwhile, but for a good high yielding variety grown for export and close to a source of fertilizer it is worth using the extra nutrients and controlling pests to achieve the maximum return of rice/hectare.

Symptoms of deficiencies of major and minor elements are well described elsewhere.

Akagare

Three different types of akagare have been distinguished in Japan. Similar diseases occur in other countries.

General methods of prevention are use of non-sulphate fertilizer, application of red upland soil and phosphates, avoidance of green manuring, use of resistant varieties and improvement of cultivation and irrigation methods.

Type 1

This occurs in ill-drained sandy lowland soils, badly drained mucks and boggy soils with a shortage of potassium. Red brown spots appear on the older leaves at the tips first, the leaves gradually die back from the tips. The plant's roots turn brown and may even go black and rot. Application of potassium fertilizers is recommended for prevention of akagare. Improved subsoil drainage may also aid prevention.

Type 2

This occurs in ill-drained muck and boggy lowland soils.

The midribs turn yellow and then red-brown spots spread over the leaves; in an acute case the spots may appear before the yellowing. Roots turn red-brown and later dark brown to black and then become rotten.

Type 2 is caused by a combination of various soil factors including excessive ferrous iron, a high organic matter content, H_2S production and low potassium. Recent research has shown that Type 2 is caused by zinc deficiency. Potassium application alone will not prevent Type 2.

Type 3

This occurs in reddish heavy clay loams, volcanic ash or upland soils which are converted to lowland rice growing systems. These soils are acidic and deficient in phosphorus.

Small brown spots appear on the tips of the lower leaves, these spread until the whole leaf becomes yellowish-brown and dies. This type of Akagare may very easily be mistaken for an attack of blast.

Phosphorus application reduces the disease. Potassium application reduces it to some extent. Recent research has shown that Type 3 is caused by iodine toxicity.

Akiochi (autumn decline)

Akiochi occurs in degraded soils typical of well drained sandy fields. The disease also occurs in poorly drained soils with a great deal of organic matter which decomposes to produce H_2S in the warm summer months.

The plants seem normal in the early stages of growth but begin to decline later in the year just before heading.

The presence of hydrogen sulphide in the soil, inhibits the uptake of potassium, silicon, magnesium and manganese, excess acids in peaty soils have the same effect. The plants have plenty of tillers but few panicles. The culms and panicles are very short, few spikelets are produced and they are either sterile, or contain small light grain with dirty brown spots on their surface. The roots are injured and the lower leaves die back. The most typical symptom of all is an extreme susceptibility to brown spot disease. The plants' resistance to the pathogen is lowered by the lack of available nutrients.

Improved soil drainage aids in control. Resistant varieties have been developed for some areas and these should be grown. Applications of basic slag which contains calcium, magnesium, manganese and silicon are very useful; as are split applications of available nitrogen and potassium. Fertilizers with sulphate radicals should be avoided.

Bronzing

Bronzing occurs in poorly drained soils in wetter areas of Sri Lanka.

During the first nine weeks after transplanting many small brown spots are seen on the older leaf tips, these may spread over the entire leaf and even onto the upper leaves in extreme cases. There are two types of soils which produce differing symptoms: 1. On sandy soil very soon after transplanting red-brown spots begin to appear. 2. On peaty soils especially if the fields are top-dressed with ammonium sulphate, yellow bronzing may occur up to two months after transplanting.

In severe cases plant growth is retarded, and there is a large number of sterile spikelets. The roots become scanty, coarse and dark brown, and their system is damaged.

Acid soil pH makes the symptoms worse and increases crop losses. The most severe cases are seen where there is hydrogen sulphide around the plant roots, and a high ferrous iron concentration in the soil.

Better field drainage, and the addition of fully ripened straw compost in sandy areas are good preventative methods. It is better to delay the flooding of fields as long as possible and to use lime and urea instead of ammonium sulphate fertilizer. Potash and sodium nitrate are helpful, the latter retards soil reduction. Heavy applications of lime were found to be effective.

Straighthead

Straighthead occurs when rice is grown in fields newly converted from grass. The disease has been reported from USA, Japan, Portugal and Colombia and is related to poor drainage on soils with a high organic matter content.

The leaves and stems become dark green and stiff and the plants are dwarfed, often with an abnormally large number of leaves. The grain weight is so low that the panicles remain upright at maturity. The hulls are distorted into typical crescent or 'parrot-beak' shapes and either the palea or lemma or both may be missing, pistils and stamens may be missing and in severe cases the entire flower. The rachis branches are thin and in extreme cases the panicles do not even emerge. Late maturing varieties are more susceptible but some resistant types have been developed. Heavy rainfall can increase the severity of the disease.

Draining the water from fields about 35 days before heading i.e. about 5—10 days before panicle differentiation depending on the variety is a good preventative measure. Using fertilizers with no sulphate radical is helpful, but the best method is to grow resistant varieties.

This is not by any means a complete list of physiological diseases. There are several others caused by specific soil conditions in many areas e.g. aogare in Japan, pansuk in Pakistan, amyit-po in Burma and khaira disease in India. Most departments of agriculture have information on local fertilizer requirements and should be consulted when plants show unexplained symptoms.

Bibliography

BABA, I. and HARADA, T. (1954). Physiological diseases of rice plants in Japan. *International Rice Commission Working Party on Rice Breeding, 5th Meeting, Tokyo.* FAO, Rome, 1954.

INADA, K. (1965). Studies on bronzing disease of rice plant in Ceylon, I—II. *Proceedings, Crop Science Society of Japan* 33 (4): 309—323.

OTA, Y. and YAMADA, N. (1962). Physiological study on bronzing of rice plant in Ceylon. Preliminary report. *Proceedings, Crop Science Society of Japan* 31 (1): 90—97.

OU, S. H. (1972). *Rice Diseases.* pp. 368. Commonwealth Agricultural Bureaux, Farnham Royal, UK.

PONNAMPERUMA, F. N. (1958). Lime as a remedy for physiological disease of rice associated with excess iron. *International Rice Commission Newsletter* 7 (1): 10—13.

SHIRATORI, K., SUZUKI, T. and MIYOSHI, H. (1969). On a disease of rice plant similar to 'Akagare' occurring in paddy field where soil dredged from the Tone River is used. *Bulletin of the Chilea Agricultural Experiment Station* 9: 73—81.

TANAKA, A., SHIMONO, K. and ISHIZUKA, Y. (1968). On 'Akagare' of lowland rice caused by zinc deficiency. *Journal of the Science of Soil and Manure, Tokyo* 40: 415.

TANAKA, A. and YOSHIDA, S. (1970). Nutritional disorders of rice in Asia. *Technical Bulletin International Rice Research Institute* No. 10.

WATANABE, I. and TENSHO, K. (1970). Further study on iodine toxicity in relation to 'reclamation akagare' disease of lowland rice. *Soil Science and Plant Nutrition* 16: 192–194.

NEMATODES

Introduction

The association of nematodes with some pathological and physiological maladies of rice has only recently been fully appreciated. Nematodes pierce the cell walls of the plant tissues by means of a hollow tubular spear or stylet, so gaining access to plant cell contents, and often injecting saliva into the host tissues during this process. The wounds or lesions caused by nematodes weaken plant tissues and also serve as foci for the entry of fungal and bacterial pathogens. The symptoms of nematode infection differ with the nematodes feeding site, the stage of a development of the infected plant or tissue, the reaction of the host plant, the density of the nematode population and the extent of the secondary infection by other pathogens. Most plant parasitic nematodes have a soil phase in their life cycles. Estimates of quantitative losses to plants and crop produce suggest the necessity for the development of cultural and chemical methods to control nematodes, particularly in the earlier stages of plant growth.

The use of nematicides in fields may be uneconomic, in Thailand it has been shown that useful economic returns can be achieved by treating the seed beds only. Cultural practices such as stubble burning are widely used for nematode control. Crop rotation will also reduce nematode infestation.

LEAF AND STEM NEMATODES

Aphelenchoides besseyi

Rice leaf nematode, or white tip disease

Distribution

Bangladesh, Chad, Dahomey, El Salvador, Gabon, Ghana, India, Italy, Ivory Coast, Japan, Kenya, Malagasy Rep., Nigeria, Senegal, Sierra Leone, Taiwan, Tanzania, Togo, southern USA, USSR.

Symptoms

The nematode is ectoparasitic, feeding on young and unfurled leaves (Fig. 33) This results in the leaves withering from the tip downwards, giving a whip-like or whitish appearance to their tips during the wet season. Chlorotic patches are also seen on the leaves. In the dry season the leaves assume a bronze colour. Before heading the leaf margins are often curled, especially those of the flag leaf. Infected plants also become stunted and tillering is reduced. These symptoms

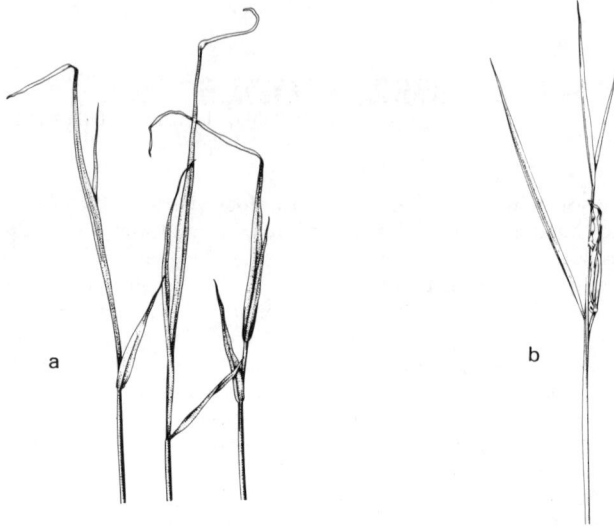

Fig. 33. Symptoms of injury due to *Aphelenchoides besseyi;* seedlings showing typical 'white tip' (a) and leaf emerging twisted (b). (H.R.B.)

have been confused with calcium and magnesium deficiency in the USA. At the heading stage, the flag leaf emerges twisted, earhead length is reduced and some glumes are rendered partially or entirely empty. Not all grains in one panicle are attacked, there may be misshapen or sterile grains alongside the normal ones.

Life cycle

At the ear formation stage of the plants, the nematodes migrate to the glumes and remain coiled up inside the palea of the grain. These nematodes are carried with the seed and can remain viable for as long as three years.

The nematodes are activated at 20°C in wet seed and are positively attracted to the young growing parts, expressed juices or aqueous extracts of succulent parts of rice plants. Symptoms of nematode injury appear in the four weeks following infection. The degree of chaffiness of the grain may be as high as 40% in certain varieties, but early ripening varieties are relatively less affected. Some varieties are resistant and the nematodes developing in these varieties are smaller than normal.

Control

A. besseyi is carried over from crop to crop in seed or chaff, it cannot survive in the soil. Burning the stubble etc. will, therefore, reduce infestation in a field. It is dispersed from infected seed by the irrigation water. In Japan, *Cyperus iria* is an alternative host. The presence of 200–300 nematodes per hundred seeds warrants control measures being taken. The treatments shown in Table 3 have been found satisfactory.

Nematicides

Rice seedlings may be treated by dipping in a thiabendazole solution for 24 h (use manufacturers recommendation for concentration).

The soil may be treated with granular or liquid formulations of nematicides. Carbofuran, cartap, diazinon, disulfoton, fensulfothion and phorate have all been used as soil treatments against this nematode.

Hot water seed treatment

Small amounts of seed should be held in water at 55–61°C for 10 to 15 min. Large quantities of seed should be pre-soaked for 24 h and then treated for 15 min at 51–53°C.

Resistant varieties

Some varieties of rice are resistant to white tip disease. In USA, Arkansas Fortuna, Asahi, Bluebonnet, Bluebonnet 50, Improved Bluebonnet, Century 231, Fortuna, Hill long grain, Nira, Nira 43, Rexoro, Sunbonnet, Texas Patna, Toro, TP-49 are resistant and Century 52, Century Patna 231 and Rexark are fairly resistant. The Japanese varieties Norin 8 and Norin 43 are resistant and Norin Mochi 5 and Natsushimo are fairly resistant. Gumartia and Chinoor in India are almost immune and Surmatia is hardly affected.

TABLE 3. NEMATICIDE SEED TREATMENTS FOR CONTROL OF
APHELENCHOIDES BESSEYI

Nematicide	Dilution in water	Time
Seed soaks		
Cartap	1:1000	24 h
Diazinon 40% e.c.	1:1000	24 h
Phosphamidon 0.1%–0.25% a.i.		24 h
REE 40% e.c.	1:500	24 h
REE 40% e.c. plus one		
ethyl mercury tablet 0.5% as Hg	1:250	12 h
Fenitrothion 50% e.c.	1:1000	24 h
Fenthion 50% e.c.	1:1000	24 h
Thiabendazole 75 0.1–0.3%		24 h
Dusts		
DSP mixed with the seed at 10% by weight		

Ditylenchus angustus

Rice stem nematode

Distribution

Bangladesh, Burma, India, Malagasy Rep., Malaysia, Philippines, Thailand, United Arab Republic.

Symptoms

The nematodes feed ectoparasitically on the tissue of unfurled leaves and sheaths, growing buds and ear primordia (Fig. 34). Injury results in yellowish or whitish-green splash patterns on the affected areas of the leaves and sheaths, whose margins emerge corrugated. The collective symptoms are known as Ufra disease. Nematodes collect between the inner side of the leaf sheaths and the unemerged developing inflorescences. The inflorescences emerge crinkled with empty glumes or do not emerge at all, unlike white tip disease when the inflorescence may be partially filled.

The symptoms of injury appear within one week in young plants and in ten to fifteen days in plants at, or prior to flowering. Brown spots due to nematode punctures may be seen on the ear sheath. The optimum temperature for infection is 28—30°C.

Fig. 34. Damage to inflorescence caused by *Ditylenchus angustus.* (J. Bridge)

Life cycle

This nematode species feeds exclusively on cultivated and wild rice varieties at all stages of plant growth. Water and high humidity are essential for their development and dissemination. In the absence of a rice crop, *D. angustus* survives in self-sown rice plants, ratoons and in stubble left in the fields after harvest. The nematodes coil up in the pre-adult stage into a woolly mass in the stubble and on the soil surface and are activated again when the fields are flooded. Seedlings in the seedbeds are first infected by nematodes carried in irrigation water flowing from the fields where they have survived between crops. In the transplanted crop, the infection is introduced by the infested seedlings, and then spread by the irrigation water, thus bringing about a high disease incidence. A laboratory test is necessary to distinguish between specimens of *D. angustus* and *A. besseyi.*

Control

Drying out the fields, burning stubble and destroying ratoon rice between rice crops are recommended. The rotation of rice with jute has been recommended in India. The use of D-D, DBCP and methyl bromide in the field or seedbed is effective. Spraying the soil with diazinon (100 ppm) where rice is planted is recorded as controlling these nematodes within 17 h.

ROOT NEMATODES

Meloidogyne spp.
M. incognita, M. graminicola, M. javanica

Root knot nematodes

Symptoms

Leaves of affected plants become discoloured from 10—12 days after the invasion of their roots by the infective second stage larvae. Leaves or leaf tips dry up, or change to a bronze colour from the margins towards the mid-rib. Occasionally, the leaf margins become corrugated. Severe infestation results in a reduction of plant height and a frayed appearance of the tillers (Fig. 35). In certain varieties, symptoms of blast disease also appear on the leaves.

On the roots distinct galls shaped like spindles, beads or clubs develop where the nematode takes up its feeding position in the root, its stylet in the stele. Profuse development of extra side roots takes place, apparently to compensate for the loss of the affected roots. The infected portion of the root becomes curved, and on the convex side of this curvature a number of small slender roots develop (Fig. 36).

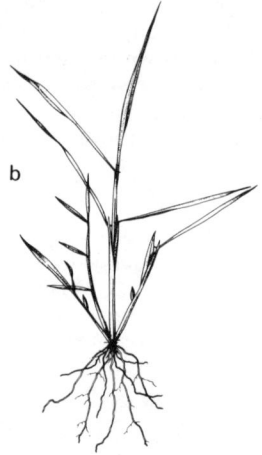

Fig. 35. Symptoms of foliar injury due to *Meloidogyne incognita* var. *acrita;* healthy plant 40 days old (a) and infested seedling 40 days old (b). Note reduced height and frayed appearance of tillers. (H.R.B.)

Fig. 36. Symptoms of root injury due to *Meloidogyne incognita* var. *acrita.* Root systems about 40 days old showing galls and development of extra side roots. Note curling of affected roots. (Central Rice Research Institute)

Under well-drained upland conditions, nurseries and newly planted crops are usually severely affected. Ill-drained or water-logged soils are not conducive to infections. Uprooting seedlings from nursery beds breaks the roots above the galls so that the galls remain buried in the soil.

Life cycle

The second stage larvae enter young roots from the soil and begin to feed in the pericycle: the cells surrounding the head become giant cells and disrupt the vascular bundles. This mechanical disruption of the conducting tissue interrupts water uptake by seedlings in the early stages and perhaps also affects their mineral nutrition at later stages of growth. A plant may have up to 1,850 galls in its root system. The mature nematodes lay eggs which are extruded into a gelatinous matrix often projecting out of the root into the soil.

Control

Flooding the soil, destroying alternate or collateral weed hosts or bare fallow periods reduce the population of these nematodes in the soil. Crop rotation is unlikely to be effective owing to the wide host range of root knot nematodes. Distinct varietal reactions have been observed in rice and varieties resistant to some species of root knot nematodes have been recognised.

Injecting seedbeds with D-D mixture or with DBCP two weeks before sowing is effective in reducing root knot, producing vigorous growth of seedlings, and increasing green weight and root development.

Heterodera oryzae

Rice cyst nematode

Distribution

India, Ivory Coast, Japan.

Symptoms

The chief symptom of infestation by this nematode is a retardation in the growth of rice plants, a depletion of vigour, and chlorosis of their leaves. The symptoms are not readily detected until over 100 cysts have developed on the root system. In severe cases, the emergence of the inflorescence is delayed.

Since the affected plants bear very few root hairs, the efficient uptake of nutrients particularly iron and nitrogen is reduced. This results in a reduction in plant size and yield. *Heterodera sacchari* has been recognised as parasitising rice and probably behaves in the same way as *H. oryzae*.

Life Cycle

The second stage larvae enter roots in the zone of elongation, penetrate the cortex, align themselves at right angles to the stele and start to feed on cells

101

around the head. As the nematode grows, the affected roots turn brown and root hairs are reduced. The cysts are lemon shaped and appear white at first; later they turn through deep brown to black.

Usually plants in upland and well-drained rice fields are affected. Infestation is highest in dry fields which are direct sown mechanically and which are flooded 3—4 weeks later. Transplanted crops are less affected than crops raised by the direct sowing method. No rice variety resistant to this nematode has been isolated.

Control

Rotating rice with non-hosts of this nematode helps to reduce infection, but cysts 2 years old and more can still produce nematodes in significant numbers. In Japan rice grown on infested land which had not been used for rice for 3 years produced 2—3 times the yield of infested land on which rice had been grown for the previous 3 years.

Pre-planting applications of D-D at 300 l/ha increased yield by 30% on infested land in Japan.

Hirschmanniella spp.
H. mucronata   H. oryzae
H. imamuri   H. spinacaudata
Rice root or burrowing root nematodes

Symptoms

Plants infected by these nematodes gradually turn yellow and wither and there is a reduction in plant height, but these symptoms are not manifest until a large population of nematodes has built up in and around rice plant roots.

The affected roots first turn brown due to the migration and feeding of larvae and adult nematodes in the cortex. The cortical cells lose rigidity and roots collapse like hollow tubes.

More nematodes are found in ill-drained fields than in well-drained fields. Infestation takes place in nursery soils and a high population can cause reductions in seedling root development and plant height.

Life cycle

The population of nematodes builds up to a maximum at the flowering time of the plant. There is an indication of a positive correlation between the level of application of nitrogenous fertilizer and the population density of the nematodes, while with potash or compost fertilizers, the population is kept at a low level. A positive correlation has been found between the nematode population and the pH of soil.

102

Several weeds in and around rice fields harbour the nematodes. Thus during the off-season, the nematodes can exist on the roots of weeds, self-sown rice, dead stubble, and can even survive in soil without multiplication.

No differences in the degree of susceptibility among rice varieties has been found.

Control

Experiments with a soil conditioner, HSC, containing 3% by weight of *Arthrobotrys* sp. a nematode trapping fungus indicated that 60 kg HSC/ha would give a yield increase. *Hirschmanniella* spp. decreased in number during the experiments. Neem and mustard cake applied to infested fields reduced the nematode population in India.

Fumigating the soil with D-D or DBCP will destroy nematodes and lead to yield increases. Fumigation should be done about 2 weeks before planting. In India drenching the soil, following irrigation, to 15—20 cm deep with thionazin at 12.5 l/ha was effective for 60 days.

Tylenchorhynchus martini

Stunt nematode

Symptoms

These nematodes are ectoparasites feeding on the epidermal cortical cells of rice roots. Large numbers of them are found in the soil around the roots. There is a general reduction in plant height, but no perceptible retardation of the root system.

Infestation is severe in low-lying rice fields where successive rice crops are raised without either a fallow or a rotation crop. Infestation is low in the nurseries.

Control

Fumigation of soils with 45% methyl bromide, EDB, D-D and DBCP a month before planting is reported to raise yields significantly, all treatments decreased the number of nematodes.

Macroposthonia onoensis

Ring nematode

Symptoms

These nematodes are ectoparasites of rice roots. They can cause severe stubby root, root swellings, stunting and chlorosis of rice seedlings. Losses can be as high as 50% with large numbers of nematodes, and it was estimated that this species caused a 15% loss in Louisiana rice production in 1967. The greatest damage is caused early in crop development.

The common rice field weeds *Echinochloa colonum*, *Cyperus* spp., and *Fuirena* sp. are hosts for the nematodes.

Control

The nematodes are eliminated by flooding of rice fields and are only likely to be a problem in seedlings and upland rice. Preplant application of fensulfothion at 67 kg a.i./ha controls *M. onoensis*.

Bibliography

ATKINS, J. G., FIELDING, M. J. and HOLLIS, J. P. (1957). Preliminary studies on root parasitic nematodes of rice in Texas and Louisiana. *FAO Plant Protection Bulletin* 5 (4): 53–56.

BUANGSUWON, D., TONBOON-EK, P., RUJIRACHOON, G., BRAUN, A. J. and TAYLOR, A. L. (1971). *Rice diseases and pests of Thailand.* pp. 61–67. Rice Protection Research Centre, Ministry of Agriculture, Thailand.

CENTRAL RICE RESEARCH INSTITUTE (1966, 1967). Study of plant parasitic nematodes affecting rice production in the vicinity of Cuttack, Orissa, India. *Annual reports of the PL-45 scheme.* Central Rice Research Institute, Cuttack.

DASTUR, J. P. (1936). A nematode disease of rice in the Central Provinces. *Proceedings Indian Academy of Sciences* 4: 108–121.

GOTO, K. and FUKATSU, R. (1956). Studies on the white tip of rice plant, III – Analysis of varietal resistance and its nature. *Bulletin of the National Institute of Agricultural Sciences,* series C, 6: 123–149 (In Japanese).

HASHIOKA, Y. (1964). Nematode diseases of rice in the world. *Il Riso* 13 (2): 139–147.

HOLLIS, J. P. (1969). Chemical control of soil nematodes in rice fields. *Phytopathology* 59: 1031 (Abst.).

HOLLIS, J. P. (1972). Nematicide – weeds interaction in rice fields. *Plant Disease Reporter* 56: 420–424.

HOLLIS, J. P., EMBABI, M. S. and ALHASSAN, S. A. (1968). Ring nematode disease of rice in Louisiana. *Phytopathology* 58: 728–729 (Abst.).

HOSHINO, M., ODA, K., TAKITA, Y., YANAKA, S., KUMAZAWA, T., TANAKA, S. and KEGASAWA, K. (1964). Studies on rice cyst nematodes, IV–VI. *Proceedings Kanto-tosan Plant Protection Society* 11: 109–111.

ICHINOHE, M. (1964). A review of the studies on nematodes attacking rice (Japan). *International Rice Commission, Working Party on Rice Production and Protection, 10th Meeting,* Manila, Philippines. FAO, 1964.

INOUE, N. and ITO, H. (1964). Effect of nematicides against rice root nematode. I. Effect of spring time treatment on transplanting cultivation. *Annual Report Plant Protection Society of Northern Japan* 15: 194–195.

ISHIKAWA, M. (1965). Relation between fertilization and rice root nematode. *Proceedings Kanto-tosan Plant Protection Society* 12: 115.

ISRAEL, P., RAO, Y. S. and RAO, V. N. (1966). Rice parasitic nematodes (India). *International Rice Commission Working Party on Rice Production and Protection, 11th meeting,* Louisiana. FAO, 1966.

ISRAEL, P., RAO, Y. S. and RAO, V. N. (1966). Survey of nematodes in rice fields and evaluation of their damage. *International Rice Commission Working Party on Rice Production and Protection, 11th meeting,* Louisiana. FAO, 1966.

KAWASHIMA, K. (1965). Control method of rice cyst nematode in direct sown dry paddy field. *Annual Report Plant Protection Society of Northern Japan* 16: 127.

KAWASHIMA, K. and FUJINUMA, T. (1965). On the injury to the rice plant caused by the rice root nematode (*Hirschmanniella oryzae*). *Research Bulletin, Fukushima Prefectural Agricultural Experiment Station* 1: 57—64.

KOMORI, N., KAWATA, S. and TAKANO, S. (1963). Studies on the control of the white tip of rice plant. *Bulletin Ibaraki Agricultural Experiment Station* 5: 1—14.

KUMAZAWA, T. (1965). Ecology of the rice cyst nematode and its control. *Proceedings Kanto-tosan Plant Protection Society* 12: 6—8.

LAVABRE, E. M. (1959). Note sur quelques parasites du riz recontrés au Cameroun avec mention d'une nouvelle espèce. *Riz et Riziculture* 5 (1): 37—41.

LUC, M. (1959). Nématodes parasites ou soupçonés de parasitisme envers les plantes de Madagascar. *Bulletin Institut Recherche Agronomique Madagascar* 3: 89—101.

MATHUR, V. K. and PRASAD, S. K. (1973). Control of *Hirschmanniella oryzae* associated with paddy. *Indian Journal of Nematology* 3 (1): 54—60 (published 1974).

MERNY, G. (1970). Les nématodes phytoparasites des rizières inondées en Cote d'Ivoire I. Les espèces observèes. *Cahiers ORSTOM, Série Biologique* 11: 3—43.

MILLER, P. R. (1953). Lowlands rice culture as an economic control of *Sclerotinia* and root knot nematodes. *FAO Plant Protection Bulletin* 1 (12): 183—187.

MUNIAPPAN, R. and SESHADRI, A. R. (1965). On the occurrence of white tip nematode of rice *Aphelenchoides besseyi* Christie 1942, in Madras State. *Madras Agricultural Journal* 51 (12): 510—511.

NAKAZATO, H., KAWASHIMA, K. and KUROSAWA, T. (1964). Seasonal occurrence of the rice root nematodes. *Proceedings Kanto-tosan Plant Protection Society* 11: 106.

NISHIZAWA, T., SHIMIZU, K. and NAGASHIMA, T. (1972). Chemical and cultural control of the rice cyst nematode *Heterodera oryza* Luc et Berdon Brizuela and hatching responses of the larvae to some root extracts. *Japanese Journal of Nematology* 2: 27—32.

SAMANTARAY, K. C. and DAS, S. N. (1971). Control of plant parasitic nematodes in rice soil. *Proceedings of Indian National Science Academy, Biological Sciences* 37B (5): 372—376.

SRIVASTAVA, A. S. and SAXENA, H. P. (1956). Effect of diazinon on paddy nematodes. *Proceedings 43rd Indian Science Congress,* New Delhi. Indian Science Congress, 1956. pp. 286.

TAYLOR, A. L., KAOSIRI, T., SITTACHI, T. and BUANGSUWON, D. (1966). Experiments on the effect of nematodes on the growth and yield of rice in Thailand. *FAO Plant Protection Bulletin* 14 (1): 17—25.

UEBAYASHI, Y., AMANO, T. and NAKANISHI, I. (1971). On the chemical control of the rice white tip nematode *Aphelenchoides besseyi* Christie 1942. *Bulletin of the Aichi Ken Agricultural Experiment Station* No. 25, pp. 50—70.

VUONG, H. H. (1968). Notes preliminaire sur la présence des nématodes parasites du riz à Madagascar *Aphelenchoides besseyi* (Christie) 1942, *Ditylenchus angustus* (Butler 1913) Filipjev 1936. *Agronomie Tropicale* 23: 1025—1049.

VUONG, H. H. (1970). Utilization du thiabendazole dans les traitements des semences du riz contre *Aphelenchoides besseyi,* nématode parasite du riz à Madagascar. *International Congress of Plant Protection (7th),* Paris, September 1970 — summaries of papers, pp. 191.

VUONG, H. H. and RODRIGUEZ, H. (1970). Lutte contre les nematodes du riz à Madagascar (resultats d'experimentation 1968—1969). *Agronomie Tropicale* 25 (1): 52—66.

VUONG, H. H. and RODRIGUEZ, H. (1972). II Resultats des essais du lutte chimique contre les nématodes parasites du riz à Madagascar; *Aphelenchoides besseyi* Christie 1942, *Ditylenchus angustus* (Butler 1913) Filipjev 1936. *Document, Institut de Recherches Agronomiques à Madagascar* No. 335, pp. 25.

WATANABE, T., YASUO, M., ISHII, K., NAGAI, M. and ICHIKI, K. (1963). Studies on the malnutrition in upland rice resulting from successive cropping. *Journal Central Agricultural Experiment Station* 5: 1—44.

YOKOO, T. (1971). Effect of the application to rice paddy of soil conditioner containing spores of the nematode trapping fungi, *Arthrobotrys* spp. on plant growth and nematode population. *Agricultural Bulletin of Saga University* No. 31, pp. 65—70.

MOLLUSCS

Lanistes ovum (Gastropoda: Ampullariidae)

Distribution

Central and Southern Africa. (Fig. 37)

Symptoms

Complete loss of stand can be experienced (Fig. 38) where, as an alternative to drilling seed into a dry seed bed, pre-germinated rice seed is broadcast into paddies flooded with 10—15 cm of water. In this the snails eat the emerging radicle and plumule. Attacks can be very insidious as seed may appear not to have germinated. *L. ovum* also feeds on the leaves of younger submerged plants and on tillers, causing less severe damage.

Life cycle

The adult snails can survive dry conditions between crops for at least eight months by burrowing into drying mud and sealing their shell aperture with the

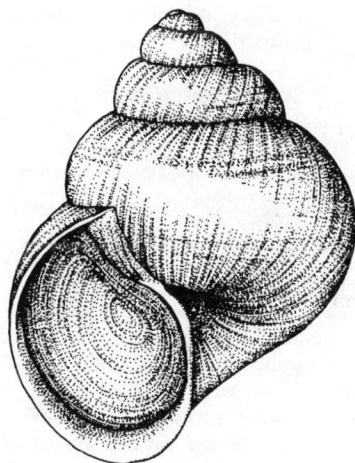

Fig. 37. *Lanistes ovum,* adult snail showing shell aperture closed by operculum. (H.R.B.)

O   2cm

Fig. 38. Aerial view of rice fields in Swaziland showing uneven stands of rice resulting from damage by snails and tadpole shrimps. (N. O. Crossland)

operculum. They are relatively long lived and after any successful attempts at controlling this pest, population resurgence could be expected to be slow.

The snails lay eggs in gelatinous masses anchored to plant stems when the rice fields are flooded to a depth of more than 15 cm.

Control

The snails are large, (up to 7 cm high) and very difficult to control, a wide range of molluscicides has been tried, but field work has shown effective control can be achieved with aerial sprays of fentin acetate at 560—1,000 g a.i./ha. Laboratory work shows that triphenyl lead acetate and triphenyl zinc chloride are as effective as the tin acetate. Unfortunately these compounds are phytotoxic to pre-germinated rice seed so that it is essential to treat fields with fentin acetate about one week before sowing to avoid chemical damage to the rice seedlings.

Ampullaria lineata (Gastropoda: Ampullariidae)

Ampullaria glauca

Distribution

Ampullaria lineata

South America, mainly Brazil.

Ampullaria glauca

South America from Bolivia to Trinidad.

Symptoms

When rice seed is sown into water in Surinam very poor emergence often results. The damage is caused by water snails, which suck newly emerged plumules from the germinating seed. Small numbers of snails can cause a great deal of damage in this way, the results of their feeding leaves large bare patches in the rice field. The worst time for an infestation of snails is during the first week after sowing, and in some cases it may be necessary to reseed after a bad snail attack.

Life cycle

During the drought before rice is planted snails creep into the mud making it difficult to recognise that fields are infested. After the field has been flooded the snails reappear and begin to feed on the newly sown rice. The snails have dark coloured striped shells and are 3—5 cm when fully grown. The two species may be distinguished by their egg clusters. *A. lineata* produces red eggs, smaller than the green ones of *A. glauca*.

Control

The snails are eaten in large numbers by the snail hawk *Rosthramus s. sociabilis* whose presence in rice fields should be encouraged. Chemical control using copper preparations and BHC (gamma and delta isomers) has been tried.

Bibliography

CROSSLAND, N. O. (1964). New rice pests in Swaziland. *World Crops* 16 (3): 51—52.

WIT, T. P. M. de (1960). The Wageningen rice project in Surinam; a study on the development of a mechanized rice farming project in the wet tropics. English Edition. Gravenhage, Mouton, 1960. pp. 293.

CRUSTACEA

SHRIMPS

Triops spp. (Branchiopoda: Notostraca)

Tadpole shrimps (Figs. 39—42)

Distribution

Triops granarius

Africa, Middle East, India, Central and Eastern Asia to the North Chinese coast. Fresh and brackish waters.

Triops longicaudatus

Western North America, through Central to South America. West Indies Galapagos Islands, Hawaii, Japan and New Caledonia. Temporary fresh waters, rice fields in California and Japan.

Symptoms

Damage does not occur where rice seed is drilled or rice transplanted but is restricted to pre-germinated rice seed broadcast into flooded paddies. The young shrimps damage this rice in the early seedling stage by nipping off plumules and radicles with their strong mandibular teeth. In severely infested paddies all seed will be completely destroyed. When seedlings are established and about 12 days old, *T. granarius* does not appear to do any further damage.

In California, there is also the problem of dislodged seedlings which are blown over after their roots have been dug up by the shrimps, resulting in very uneven stands of rice.

Life cycle

Adult shrimps are olive grey and 5.0—6.5 cm in length. Their body is covered by a shield-like carapace and they possess strong mandibles, similar to those of a chewing insect. Numerous appendages bearing leaf-like gills are attached to the abdomen. Eggs are very resistant to desiccation and temperature changes and remain viable in the soil during the winter when the fields lie fallow.

When the fields are flooded in the following spring, the eggs hatch and rapidly develop into mature females. Reddish-orange egg sacs appear along the sides of the abdomen near the carapace soon after the adults emerge. There may be more than one generation per year.

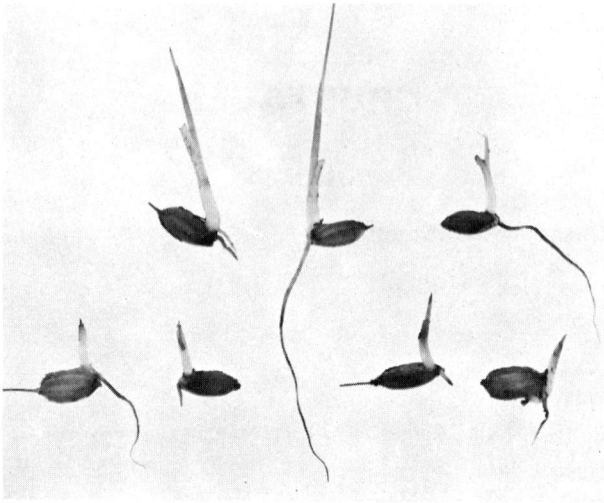

Fig. 39. Young rice seedlings damaged by *Triops*, note plumules and radicles bitten off compared with healthy seedling in the centre. (W. H. Lange)

Fig. 40. Rice on left damaged in the early stages of growth by *Triops* feeding, compared with healthy plants on the right. (W. H. Lange)

Control

The pest is easily controlled by many common insecticides. Among them aerial application of 5 g a.i. malathion/ha is both cheap and effective. DDT is effective as granules or spray applied before planting, with the seed, or after seeding.

CRABS

Two types of crabs commonly cause damage to rice, land crabs (Gecarinidae) and river crabs (Potamonidae). Crabs may also damage bunds and irrigation systems by burrowing into the banks or dams and causing leaks.

Many species of land crabs have been recorded from Asia as feeding on nursery and newly planted rice seedlings. The crabs feed mainly at night and may be controlled by the use of traps or poisoned baits, but are more often kept in check by rats, water fowl or other natural predators e.g. the paddy bird *Ardeola grayi*. River crabs have been reported from Asia, but are very serious pests in some areas of West Africa.

Sesarma huzardi (Malacostraca: Potamonidae)

River crab

Distribution

West Africa.

Symptoms

Serious losses were reported from this pest on the West Africa Rice Research Station at Rokupr, Sierra Leone, where rice is grown on tidal mangrove lands, immediately after transplanting tender seedlings are cut off at ground level, and older ones are torn apart at the base to enable the crab to reach the tender inner tissues. Crabs of several species are normally plentiful on the mangrove lands but occasionally there is an enormous increase in the population of *S. huzardi* and severe damage to the crop may be caused.

Life cycle

The crabs breed throughout the year, but mainly in the dry season and their breeding is also affected by the cycle of tides in the area. A few large eggs are laid by each female and carried after hatching within the female's flap-like abdomen for protection of the small crabs. Crabs live near the river, within the tidal area in burrows from which they emerge to feed on the rice.

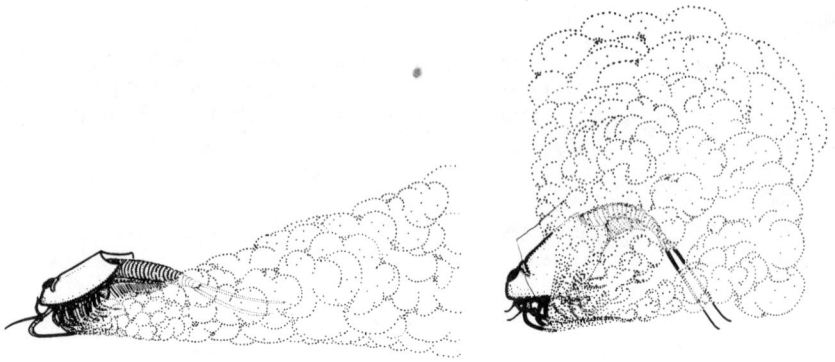

Fig. 41. Diagrams showing the digging action of *Triops* which dislodges rice seedlings. (W. H. Lange)

Fig. 42. Spraying rice from the air in Swaziland to control snails and tadpole shrimps. (N. O. Crossland)

Control

Local farmers usually do not use any chemicals to control crabs but they do plant extra seedlings in each hill of rice to compensate for any damage on plots near the river banks. Experiments have shown that crabs can be controlled by the use of sprays of BHC (6.5% gamma isomer), applied following high spring tides in the dry season. Sprays of the same BHC at 15 g/m^2 sprayed on seedlings in nurseries protected the seedlings both there and after transplanting.

In other areas different chemicals have been used as poison baits, or for poisoning burrows and spraying bunds where the crabs breed. Before undertaking control it is important to calculate how much actual loss of yield the crabs are causing.

Bibliography

BOWLING, C. C. (1964). Insect pests in the United States in *The major insect pests of the rice plant,* International Rice Research Institute. pp. 551–570. Johns Hopkins Press, Baltimore, 1967.

CROSSLAND, N. O. (1965). The pest status and control of the tadpole shrimp, *Triops granarius,* and of the snail, *Lanistes ovum,* in Swaziland rice fields. *Journal of Applied Ecology* 2 (1): 115–120.

GRIGARICK, A. A., LANGE, W. H. and FINFROCK, D. C. (1961). Control of the tadpole shrimp *Triops longicaudatus* in California rice fields. *Journal of Economic Entomology* 54 (1): 36–40.

GRIST, D. H. and LEVER, R. J. A. W. (1969). *Pests of rice.* pp. 520. Longmans, London, 1969.

JORDAN, H. D. (1957). Crabs as pests of rice on tidal swamps. *Empire Journal of Experimental Agriculture* 25 (99): 197–206.

LONGHURST, A. R. (1955). A review of the Notostraca. *Bulletin British Museum (Natural History) Zoology* 3 (1): 1–57.

INSECTS AND MITES

Introduction

In this chapter the major insect and mite pests of the rice plant are discussed under the following headings:—

Stem borers	Lepidoptera
Leaf feeders	Orthoptera
	Hemiptera
	Thysanoptera
	Lepidoptera
	Diptera
	Coleoptera
	Acarina
Stem and root feeders	Orthoptera
	Hemiptera
	Diptera
	Coleoptera

The distribution of each pest is given, where possible with reference to the relevant Commonwealth Institute of Entomology Map. These may be obtained from the Commonwealth Agricultural Bureaux, Farnham Royal, Slough SL2 3BN, England. The distribution is followed by a description of the symptoms of pest attack and an outline of the insect life cycle. Control measures conclude each section.

Before planting rice, a variety bred for resistance to the local major pest or disease should be chosen if possible. The local extension officers will be familiar with what is available. The International Rice Research Institute, P.O. Box 933, Manila, The Philippines and the Central Rice Research Institute, Cuttack-6, Orissa, India have been leading the breeding for resistance programmes in rice and should be consulted when difficulties are encountered.

There are several cultural methods of reducing pest attack. The preparation of the field for paddy destroys many of the insects present and this type of rice is attacked by insects moving in from alternative hosts or other rice fields. The majority of rice pests have alternative hosts from which insects may migrate to the rice. Destruction of weeds by reforming bunds or dykes in paddy fields, ploughing and crop rotation cuts down these alternative hosts. Immediately after harvesting, ploughing followed by flooding will effectively reduce the insect pests available to infest subsequent crops. Where the land is baked hard and ploughing cannot be done flooding should be carried out if possible. Harvested straw should be used to feed animals or composted so that it is not a source of infestation. The timing of planting to miss the egg-laying period of stem borers can also reduce infestation.

In some countries forecasting systems have been set up for particular migratory pests, e.g. in Japan for *Nilaparvata lugens* and *Sogatella furcifera* — farmers are recommended to spray when the Ministry of Agriculture considers it necessary. Some extension services or large farmers set up traps, mainly light traps but sometimes including pheromones, to assess some local pest populations. These are particularly helpful for timing the arrival of newly hatched Lepidoptera larvae since they hatch 5—7 days after the peak adult moth catches.

Once a pest insect has been found on a crop and identified its importance must be determined. In some cases it can be said that when x number of that insect are found in a sample of one hundred plants the pest will cause a loss in yield so great that it will be worth spending money on chemical control. In other cases the farmer must rely on his experience or that of his local extension officer to assess when it is necessary and profitable to apply chemical control treatments and when to expect the plants to recover without treatment or have only slightly decreased yield. Good farm records indicating the pest history of previous years, treatments and yield, are valuable for consultation under these circumstances.

When, in spite of the use of those cultural methods that are available the pest multiplies to economically important levels the choice of insecticide should be made with care as some insecticides will kill many kinds of insect including parasites and predators. The death of these beneficial insects can lead to an increased population of a pest insect — and not necessarily the one which was being treated in the first place. Insecticide manufacturers are aware of the importance of conserving beneficial insects and will sometimes indicate on their labels if an insecticide does not kill parasites and predators. When applying an insecticide the farmer should know which insect pests, apart from the one he is attempting to control, will be killed by that insecticide.

The use of costly chemical control methods pays off best on high-yielding rice varieties which are responsive to nitrogenous fertilizers. It is important to remember that the yield potential of these varieties can only be fully attained if they are grown where irrigation is controlled, adequate fertilizers are applied and pest control is practised. Such intensive conditions favour pest infestation. The use of nitrogen in particular makes rice more lush and thus prone to stem borer and leafhopper attack.

The choice of formulation, granular, e.c., dust, w.p. etc., will depend upon the formulations available and recommended locally, the money available and the spraying/dusting machinery available. Recent work has shown that one application of insecticide granules in the rooting zone (sometimes with the fertilizer) a few days after transplanting controls some insect pests for the entire crop life. This method may become very important in the future. Currently granules, though bulky and expensive, are very popular since they are easy to handle and relatively safe (see p. 263).

Whatever formulation is used safety recommendations must be followed as insecticides can be dangerous to humans and their use should be followed by washing of clothes and person. If protective clothes are required and are not available then that particular insecticide must not be used. All empty insecticide containers should be crushed or burnt. Pesticide containers should never be re-used for any purpose. Where fish are reared in the irrigation water insecticides that are toxic to fish must be avoided.

When chemical measures are thought to be necessary in an area where pest control has not been used previously, it is very important to make sure that local trials are carried out before products are recommended. This will help to ensure that the dosage used is effective and that the cost and product availability are assessed.

Throughout the sections on particular pests the insecticides are referred to by accepted common chemical names. At the end of this manual the common chemical names are listed with some corresponding trade names. This list is not complete, new trade names appear regularly and it is not possible to publish a comprehensive list. The inclusion of a trade name does not indicate approval of that product. Usually the recommendation is given as active ingredient per hectare (a.i./ha); where the formulation is a mixture of two chemicals the recommendation refers to the formulation. The recommendations in this manual are taken from many sources and different countries. Local recommendations and manufacturers instructions, if available, must be used in preference. In some countries certain insecticides are banned and therefore not available and in rice importing countries there are regulations concerning pesticide residues on rice and exporting rice growers should consider these before applying insecticides.

STEM BORERS

Lepidoptera

Lepidopterous rice stem borers are important in many rice growing areas. These pests belong to the Lepidoptera families Pyralidae and Noctuidae. The symptoms resulting from attack by all species of stem borer are similar and are therefore described before considering individual species and control.

Symptoms

Injury to the rice plant is caused by larvae tunnelling in the stems and feeding on the soft tissues. Adult stem borer moths lay eggs in masses, which are easily seen, on the leaf surfaces, except for *Sesamia inferens* which deposits its eggs between the leaf sheath and the stem. Newly hatched larvae enter leaf sheaths and begin feeding on the inner tissues. Attacked leaf sheaths first show transparent patches, later turn yellowish-brown and eventually become dry. After a few days, larvae bore into the stems and feed on stem tissue until they reach a node. They then feed on the tissues around the node, as a result of which infested stems are weakened and are easily broken. Some borers eat through the nodes and bore right down the stem; others leave the stem above the node and either enter another tiller or a neighbouring internode of the same stem.

119

Fig. 43. A white head of empty grains—
a typical symptom of stem borer
infested plants which survive to bear
panicles. (L.T. Kok)

With a few borers, notably the *Sesamia* species pupation may take place between the leaf sheath and the stem.

Visible symptoms on affected plants vary with the stage at which infestation of the plant commences. The first clear sign is the curling of the youngest seedling leaf or the presence of egg batches on the upper or lower leaf surfaces. Often adult moths are seen in the vicinity of the young plants. Young seedlings which have been attacked at the base of the stem have 'dead hearts' caused by the drying up of the tips of the central shoots. Older infested plants, at the maximum vegetative stage, have holes where the larvae have entered and disintegrated tissues and broken stems caused by their feeding. Frass at the feeding site is common where recent feeding has occurred. Plants which have started bearing panicles at the time they were attacked usually give rise to 'white heads' (Fig. 43) or discoloured panicles with empty or partially filled grains. When *Maliarpha separatella* is responsible the panicle bears some green and some white spikelets. Late attack can kill the panicle even after grain formation has started. Plants that are attacked after maximum tillering has taken place suffer most from the attack (Plate 1).

Chilo agamemnon (Lepidoptera: Pyralidae)
Small purple-lined borer

Distribution

Egypt, Iraq, Sudan, Uganda (CIE Map 299).

Control

Cultural

Since this pest is known to hibernate as a larva in the stubble a proportion can be killed by destroying stubble after harvest.

Chemical

See Table 4.

Chilo auricilius (Lepidoptera: Pyralidae)
Gold-fringed rice borer

Distribution

Bangladesh, Bhutan, Borneo, Burma, China, Hong Kong, India, Indonesia, Kashmir, Malaysia, Nepal, Philippines, Sikkim, Vietnam (CIE Map 300).

Life cycle

This species is a serious sugarcane pest in India and Taiwan and a major pest of rice in the Orissa area and hill areas of Kalimpong, West Bengal. Eggs are laid on the undersurface of leaves and hatch about a week later. The larvae, which have only four stripes (the dorsal stripe is absent), mature in 30—40 days. Pupation takes place in the affected stems.

Control

As for other *Chilo* species.

Chilo partellus (Lepidoptera: Pyralidae)
Maize or sorghum stem borer

Distribution

Afghanistan, Comoro Is., India, Kenya, Malawi, Nepal, Pakistan, Sikkim, Sri Lanka, Sudan, Tanzania, Thailand, Uganda (CIE Map 184).

Life cycle

This species is polyphagous and has been recorded as a pest of a number of cereals as well as rice. As a rice pest it is particularly important in eastern India. Eggs are laid in two 13 mm overlapping rows on any part of the plant. In 4 to 8 days the larvae emerge, these are dirty white in colour with the primary setae borne on each segment on yellowish verrucae. There are two dorsal stripes which may merge to form a wide band and lateral stripes. Newly hatched larvae are attracted to light, on maize they are sometimes found in groups on the tassel. About an hour after hatching they bore into the stem.

121

The length of the larval period depends upon the season and may be 18—193 days. Pupation takes place within the stem and lasts 6—12 days.

The adult emerges through an exit made by the full grown larva. During the winter (possibly when the temperature is below 25°C) older larvae remain dormant, some in the stems and stubble.

Control

Chemical

See Table 4;

TABLE 4. RECOMMENDATIONS FOR CHEMICAL
CONTROL OF STEM BORERS

Insecticide	Formulation	Active ingredient per hectare	Application
Acephate[2]	e.c.	0.4—0.5 kg	—
Azinphos-methyl[2,3,5,7]	—	0.3—0.5 kg	—
BHC[1,2,3,5,*]	gran.	1.8—2.4 kg	The irrigation water should be present (Resistance to BHC recorded in Japan)
Carbofuran[2,3,7,5]	gran.	0.45—0.6 kg	Apply 30 days after germination or when borer population warrants it
Cartap[2]	gran. microgran.	0.3—0.8 kg	—
Chlordimeform[2,5,7]	jumbo gran., gran. dust, e.c., s.p.	0.5—1 kg	Apply direct to water 5—20 cm deep 2—4 treatments, when moths are on the increase, at 3—4 wk intervals
Chlorfenvinphos[2,5,6]	gran.	0.2—0.3 kg	—
	e.c.	0.2—0.5 kg	—
Chlorpyrifos[1,5]	follow manufacturers recommendations		—
Cyanofenphos[2]	—	0.3—0.5%	—
DDT[2,3,5]	dust	0.1 kg	—
Diazinon[1,2,4,6]	e.c., w.p.	0.3—0.6 l	—
	gran.	1—1.5 kg	Direct to water when moths are on the increase
Dicrotophos[2,5,7]	—	0.3—0.6 kg	—
Disulfoton[4]	—	1.6 kg	—
Fenitrothion[2,7]	u.l.v.	0.6 l	Resistance to fenitrothion recorded in Japan

TABLE 4—continued

Insecticide	Formulation	Active ingredients per hectare	Application
Fenitrothion + malathion[5]	e.c.	1.0 kg (formulation)	–
Fensulfothion	gran.	1.5–2 kg	–
Fenthion[2,4,5,7]	dust, u.l.v.	0.6–1.2 kg	The larger amount for the second generation
Formothion[2]	–	0.34 kg	–
Malathion[5]	u.l.v., e.c.	1.5 kg	–
Mephosfolan[2,3,5,7]	gran.	0.5–1 kg	–
Monocrotophos[2,5,7]	gran.	0.5–1.5 kg	–
Phosmet[2,5]	follow local label recommendations		
Phosphamidon[2,5]	–	0.2–0.5 kg	–
Pyridaphenthion[2]	dust, e.c.	0.3–0.4 kg	–
Salithion[2]	gran. microgran.	1–2 kg	–
Trichlorphon[2,5,6]	u.l.v.	0.8 l	–
	w.p.	1–1.8 kg	–

* Does not control pink borers.

1 2 3 4 5 6 7 Indicates that the reference from which the recommendation was taken referred to a particular species of borer as follows [1] *C. agamemnon* [2] *C. suppressalis* [3] *C. polychrysus* [4] *M. separatella* [5] *T. incertulas* and *T. innotata* [6] *S. calamistis* [7] *S. inferens*.

Chilo plejadellus (Lepidoptera: Pyralidae)
American rice stem borer

Distribution

Mexico, USA (Louisiana, Georgia).

Life cycle

Whitish eggs are laid in groups of 50 or more overlapping each other. The yellowish white larvae have brown lateral stripes and when full grown are about 25 mm long. Pupation takes place in the spring, the larvae having spent the cool season in the stubble. Adults have light straw coloured forewings with a few black dots on them and the hindwings are white.

Control

Cultural

Serious infestations are sporadic. Cultural practices such as stubble destruction should keep damage to a minimum.

Chilo polychrysus (Lepidoptera: Pyralidae)
Dark-headed striped borer

Distribution

Burma, India, Laos, Malaysia, Philippines, Sabah, Thailand, Vietnam.

Life cycle

This species is the most important borer in Malaysia and is sometimes referred to as the paddy borer of Malaya. The adult moth is light yellow with 6—7 tiny black dots in the centre of the forewings (Fig. 44a). The hindwings are yellowish-white. It has a body length, of 10—13 mm and wing-span of 17—23 mm. It usually lives for 2—5 days.

Eggs are laid in batches of 20—150 in rows and are scale-like in appearance. They hatch in 4—7 days. The newly hatched larva feeds actively and reaches

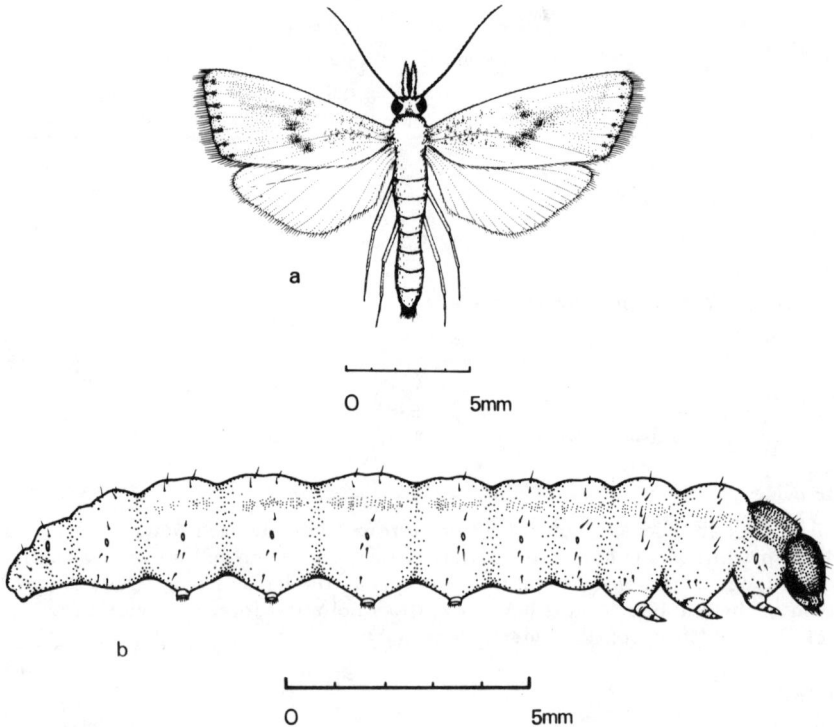

a

O   5mm

b

O   5mm

Fig. 44. Dark-headed rice borer, *Chilo polychrysus* (a) adult, (b) larva. Note black head capsule. (H.R.B.)

124

full size in about 30 days, moulting five times before pupating. The mature larva is 18—24 mm long, 2.4 mm wide and has a black head capsule and thoracic plate (Fig. 44b). On the abdomen 3 dorsal and 2 lateral purplish-brown stripes are distinct. The pupa is yellowish-brown with distinct abdominal stripes, 10 mm long and 2 mm wide. The adults emerge in 6 days. The total life cycle is 26—61 days. The presence of one larva per 4 tillers at 65 days after sowing has been shown to lead to significant yield losses.

Control

Chemical

See Table 4.

Chilo suppressalis (Lepidoptera: Pyralidae)
Asiatic rice borer, pale-headed striped rice borer

Distribution

Australia, Bangladesh, Burma, Cambodia, China, Hawaiian Is., India, Indonesia, Italy, Japan, Korea, Malaysia, Nepal, Philippines, Spain, Sri Lanka, Taiwan, Thailand, Vietnam (CIE Map 254).

Life cycle

The moth is similar to the dark-headed borer in colour but is slightly larger in size. It is 13 mm long with a wing-span of 20—23 mm although the female may reach 35 mm. The colour is straw to light brown and there are a number of silvery scales and usually five black dots in the middle of the forewing, the hindwing is yellowish-white. The moth lives for 3—5 days.

Eggs are laid on the basal half of the leaves. They resemble those of the dark-headed borer and hatch in 5—6 days. The larva when fully grown is approximately 26 mm long and 2.5 mm wide. It has a yellowish-brown head and has three dorsal and two lateral brownish abdominal stripes, the middle dorsal stripe is lighter in colour than the two along each side. It is fully grown in 33 days and changes into a reddish-brown pupa, 11—13.5 mm long and 2.5 mm wide. Moth emergence takes place after 6 days. The life cycle is 41—70 days in duration. In the cooler regions one to two generations a year are produced and full grown larvae diapause during the cool season. In the tropics six or more generations per year are possible.

The species has in some areas decreased in importance since the introduction of granular pesticides. It chiefly attacks rice in the middle or late stage of growth.

Control

Cultural

By cutting the rice as close to ground level as possible during harvesting the majority of larvae may be removed from the field, at this time they are 10—17 cm above ground level. This operation significantly reduces the infestation in the following rice crop. Stubble should be ploughed in or flooded soon after harvesting to destroy any larvae or pupae which remain.

A variety of rice which has some resistance to this stem borer should be chosen if available. Work done in Japan has indicated that direct-seeded rice is less susceptible to *C. suppressalis* than transplanted rice.

Chemical

When there are more than 10% dead hearts in the crop before 6−7 weeks after transplanting or more than 5% after then consideration should be given to the application of pesticides. It is generally recommended that insecticides are applied 4−5 days after a generation of adult *C. suppressalis* have emerged, in order to treat very young larvae.

In Spain farmers have arranged for two aerial applications of fenthion to be made over the entire rice growing area, the first as a liquid and the second as a dust, in June and July i.e. one and two months after planting. This type of cooperative arrangement can be more economical than individual action. Numerous insecticides are used for the control of this stem borer and some of these are listed in Table 4.

Maliarpha separatella (Lepidoptera: Pyralidae)
White stem borer

Distribution

Burma, China, Cameroons, Ghana, Kenya, Malagasy Republic, Malawi, Tanzania, Senegal, Sierra Leone, Swaziland, Uganda (CIE Map 271).

Life cycle

The male adult is 15 mm long, 3 mm longer than the female, both have long pale yellow wings which overlap along the body when the moth is at rest. On the paler anterior wings there is a marked red-brown line between the costal and radial veins and a brown zone along the costal vein which is more marked in the male (Fig. 45). There is a patch of brown scales on the straw coloured underside. The posterior wings are white with a metallic sheen, fringed with long hairs of the same colour.

The female, which has longer antennae than the male, lays eggs grouped in long masses on the upper surface of rice leaves. The eggs are attached to the leaf by a strong cement, as this dries it contracts, the leaf becomes puckered and the egg mass is enclosed in a foliar envelope. The eggs which are 0.6−0.7 mm long and 0.4−0.5 mm broad, are laid close together and the entire mass is normally 1.5 cm long and may contain 50 eggs. They are clear yellow when laid but become paler, and finally darken to brownish black. The larva is transparent white on hatching with a dark brown head, the pronotum has three bands of

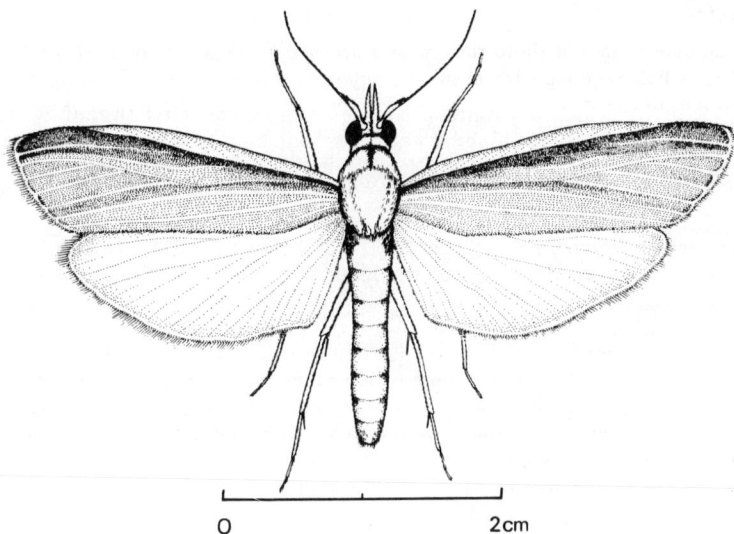

Fig. 45. White stem borer, *Maliarpha separatella*. (H.R.B.)

characteristic thickening the same colour as the head. As it grows the larva yellows. Larvae can be dispersed by the wind suspended on a silken thread.

The pupa is brown with a dorsal red spot on the fifth abdominal segment of the female, which becomes less visible as the case dries out. Pupae have been found within the stem in a loose cocoon, and they remain in the basal part of the stem during the dry season. An important character of this species is that its larvae can go into a resting stage. In Sierra Leone, for example, where there is a single rice crop and a well marked dry season, this resting stage may last up to 20 weeks on the rice stubble. The larvae then pupate, and adults emerge to lay eggs on the next rice crop. There are usually 3 or 4 generations per year.

Control

Chemical

See Table 4.

Tryporyza incertulas (Lepidoptera: Pyralidae)
Yellow stem borer, paddy stem borer

Distribution

Afghanistan, Bangladesh, Burma, Cambodia, China, Hongkong, India, Indonesia, Japan, Laos, Malaysia, Nepal, Pakistan, Philippines, Ryuku Is., Sikkim, Sri Lanka, Taiwan, Thailand, Vietnam (CIE Map 252).

127

Life cycle

The adults show sexual dimorphism, and are often mistaken for two species. The male is light brown with numerous small brownish dots, five along the subterminal area and 8—9 near the tip of the forewing. The female is yellow, the colour deepening towards the tip and there is a very distinct black spot in the centre of each forewing (Fig. 46). The hindwings are pale straw coloured. The body length is 13—16 mm and the wing span is 22—30 mm. It lives for 4—5 days.

The eggs, which hatch in 8 days are laid in oval patches of 80—150 and are covered by the brown anal hairs of the female. They are laid near the tip of the leaf blade. The yellow larva is 25 mm by 3 mm when fully grown (after 40 days), it has an orange head capsule and a rather small head relative to the width of the body. Pupae are yellowish-white with a tinge of green and measure about 16 mm long. The pupal stage which takes place inside the stem often below the soil surface, lasts for 8 days. The entire life cycle may last 52—71 days. Adults can emerge through water. This stem borer attacks throughout the growing period of the crop.

Control

Chemical

See Table 4.

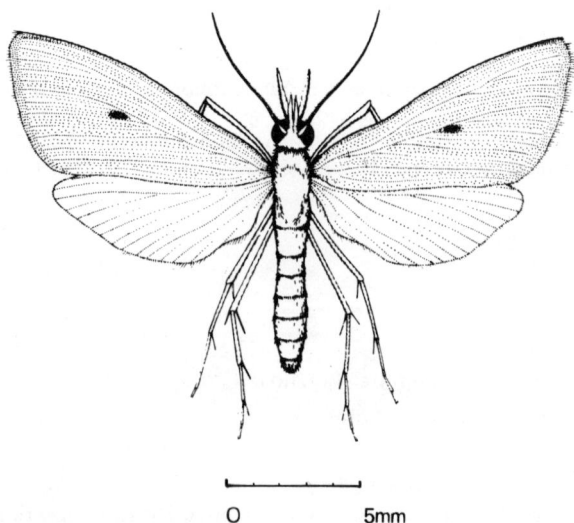

O   5mm

Fig. 46. Yellow rice borer, *Tryporyza incertulas.* (H.R.B.)

Tryporyza innotata (Lepidoptera: Pyralidae)
White rice borer, white stem borer

Distribution

Australia, India, Indonesia, Malaysia, Pakistan, Philippines, Taiwan, Vietnam (CIE Map 253).

Life cycle

The adult lives for 4—5 days, it is white, slender and similar to the yellow rice borer, but has no black spot on the forewing. The female has a pink abdominal tip.

Eggs are laid in batches of about 100 and are covered with silky greyish hairs, the incubation period being 4—9 days. The larvae are similar to those of the yellow borer except that they are white. They are fully grown in 19—31 days. The pupae are soft-bodied, pale and 12—15 mm long. Pupation is completed in 7—11 days.

Control

Cultural

When harvesting cut the straw no higher than 10 cm above ground level. Leaving a greater amount of vegetation behind is responsible for the carry-over of borers to the next crop.

Chemical

See Table 4.

Rupela albinella (Lepidoptera: Schoenobiidae)
South American white borer

Distribution

Neotropical region and nearctic: Southern USA.

Symptoms

Yellowish stained patches below the axils of leaf sheaths.

Life cycle

Conspicuous yellow clusters of eggs are laid on the leaves. After initial wind dispersal the cream coloured larvae grow to about 30 mm long, tunnelling into the nodes. Pupation takes place in the stem. The moth is a shiny white insect. The entire life cycle takes 6.5—8 weeks.

Control

Cultural

Work carried out in Surinam has indicated that chemical control for this species would be uneconomical. Stubble burning and clean cultivation of fallow fields have given efficient control.

Sesamia calamistis (Lepidoptera: Noctuidae)
Mauritius pink borer of sugarcane

Distribution

Africa.

Control

Chemical

See Table 4.

Sesamia inferens (Lepidoptera: Noctuidae)
Pink borer

Distribution

China, India, Japan, Korea, Malaysia, Pakistan, Philippines (CIE Map 237).

Life cycle

The adult (Fig. 47a) is fawn with dark brown markings. From a central point in the forewing, a typical radiation of greyish black spreads towards the wing tips, ending in a thin line of dark spots to form a terminal line. The hindwings are white. The body length is 14–17 mm and wing span can reach up to 33 mm. It lives for 4–6 days.

Bead-like eggs are laid in rows between the leaf sheath and the stem, usually from 30–100 per batch. The incubation period is 7 days. The larva (Fig. 47b) has an orange-red head capsule and its body is purplish-pink dorsally and white ventrally. When fully grown at 36 days it can reach 35 mm in length and 3 mm in width. The pupa is dark brown with a tinge of purple on the head region and is 18 mm long and 4 mm wide. The total life cycle lasts 46–83 days. This stem borer is a major pest in double cropped areas.

Control

Chemical

See Table 4.

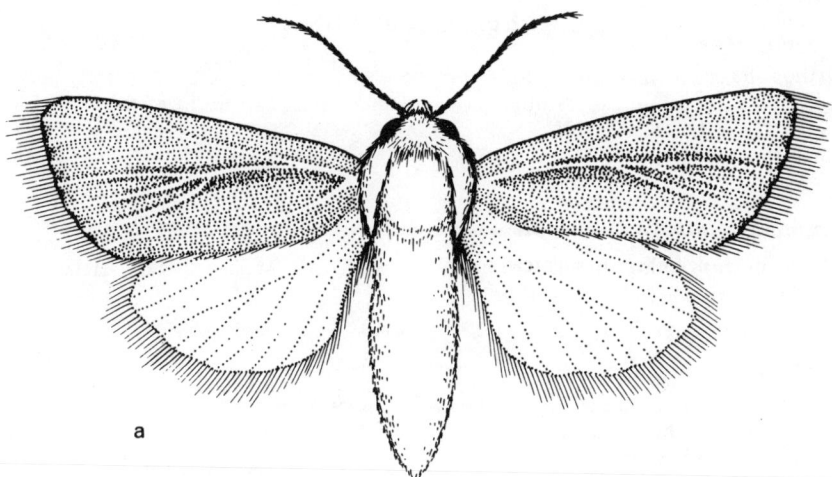

a

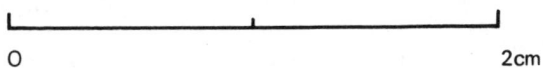

O   2cm

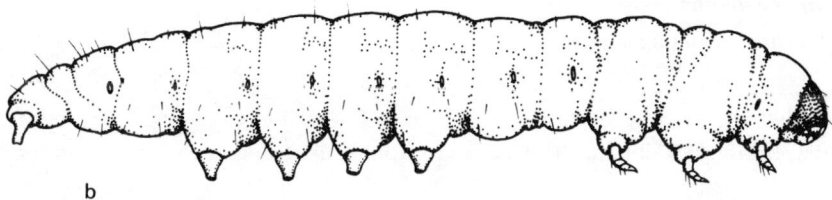

b

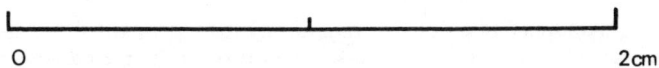

O   2cm

Fig. 47. Pink borer, *Sesamia inferens,* (a) adult, (b) larva. (H.R.B.)

LEAF FEEDERS

Orthoptera

Oxya spp. (Orthoptera: Acrididae)
Small rice grasshoppers

Distribution

Africa, Burma, China, India, Indonesia, Japan, Korea, Malagasy Rep., Malaysia, Taiwan.

Symptoms

Immature and adult insects eat the leaves of nursery plants and later in the growing season the adults attack the base of the rice panicle causing it to wither, this is often the most important attack.

Life cycle

In dry conditions eggs are laid just below the soil surface and in wet conditions among rice stems and grasses 2.5—5 cm above water level. Eggs are laid in a gummy froth which hardens to form a protective pod. In southern areas (south of the January 17.7°C isotherm) breeding is more or less continuous throughout the year, especially where both early and late varieties of rice are grown. In more northern areas there is one generation per year and the insect overwinters in the egg stage. There are usually 6 instars with many females going through an extra moult.

Hieroglyphus banian (Orthoptera: Acrididae)
H. nigrorepletus
Large rice grasshoppers

Distribution

India, Pakistan, Vietnam, Thailand, south China, Sri Lanka.

Symptoms

Like most grasshopper pests of rice the normal habitat of these insects is marshy areas and wet ditches with grassy banks, typical of most rice growing regions. The young nymphs feed on germinating seedlings in the nurseries causing withering. The adults feed on the leaves and stems of mature rice plants, often causing the panicles to wither.

Life cycle

Eggs are laid in the soil of the grassy bunds around the rice fields and the nymphs move onto the rice after they hatch. There is usually one generation per year and the adult overwinters in weed grasses around the rice fields.

Control of grasshoppers

Cultural

Ploughing after harvesting is recommended to bring the eggs to the surface and to destroy them.

Chemical

DDT, dichlorvos, fenthion, phosphamidon or diazinon will control these grasshoppers when used at the usual rates for rice.

Hemiptera: Homoptera

Nephotettix spp. (Hemiptera: Euscelidae)

Green rice leafhoppers

Life cycle

Adults are 3.2–5.3 mm long, greenish in colour with black spots on the wings. Eggs are laid in the leaf sheaths where they hatch in about 6 days. The nymphs have a varied colour pattern on the notum and they undergo five moults to reach the adult stage in 16–18 days (Plate 2).

Nymphs and adults cause direct damage by sap feeding from the leaves and base of the plants. They are also important as virus vectors (see pp. 87–89).

N. cincticeps

Vector of transitory yellowing, yellow dwarf and dwarf virus diseases.

Distribution

Japan, Korea, Manchuria, Ryuku Is., Taiwan.

Control

Cultural

Varieties resistant to this leafhopper should be used where they are available.

Chemical

See Table 5 for recommended materials. The first application should be on transplanting or when flooding.

TABLE 5. RECOMMENDATIONS FOR CHEMICAL CONTROL OF *NEPHOTETTIX* SPP.

Insecticide	Formulation	Active ingredient per hectare	Application
Acephate	gran.	0.022–0.033 kg	To the water
Acephate + carbaryl	fine gran.	follow manufacturers recommendation	Recorded controlling carbamate-resistant *N. cincticeps*
Allyxycarb	–	0.03–0.05%	–
Azinphos-methyl	–	0.3–0.5 kg	–
BHC + carbaryl	gran.	12–18 kg (formulation)	Broadcast in 3–5 cm water retain water for 7–10 days after insecticide application
BHC + isoprocarb	gran.	3–4 kg (formulation)	Into the water maintain water level 3–4 days
BPMC	u.l.v.	0.5 l	–
	gran.	1.2–1.6 kg	–
Carbaryl	follow manufacturers recommendation		
Carbofuran	gran.	0.45–0.6 kg	Into the water 15 DAT[1] repeated at 45 and 75 DAT
Chlordimeform	gran.	1.0–1.5 kg	Broadcast into 3–5 cm water. Retain the water for 7–10 days after insecticide application
Chlorpyrifos	follow manufacturers recommendation		
Cartap + BPMC*	–	0.1 kg formulation per box	For young plants grown in seed boxes for transplanting machines apply 3 days – just before transplanting
Diazinon*	–	0.085–0.1 kg formulation per box	
Disulfoton*	gran.	0.1 kg formulation per box	
Diazinon	gran.	1–1.5 kg	Broadcast in 3–5 cm water retain water for 7–10 days after insecticide application
Dicrotophos	–	0.3–0.5 kg	–
Dimethoate	follow manufacturers recommendations		
Demeton-*S*-methyl	–	0.75–1.5 kg	–
Disulfoton	gran.	1.0 kg	Broadcast at planting or prior to flooding
Endosulfan	gran.	1.0 kg	–
Fenitrothion	e.c.	1.0 kg	
Fenitrothion + malathion	u.l.v.	0.5 l (formulation)	–

TABLE 5—continued

Insecticide	Formulation	Active ingredient per hectare	Application
Fenitrothion + MPMC	—	25 kg (formulation)	—
Fensulfothion	gran.	1.5 kg	—
Fenthion	dust	0.6—1.2 kg	—
	e.c.	0.75 kg	—
Formothion + disulfoton	gran.	3.4 kg (formulation)	—
Isoprocarb	gran.	1—1.5 kg	To water surface
	dust	0.6—0.8 kg	
Isothioate	gran.	3 kg	Surface soil treatment in the nursery
Malathion	u.l.v.	0.5 l	Resistance to this insecticide is common in Japan
Mephospholan	gran.	1.2 kg	—
Methomyl	—	0.5—0.7 kg	—
Naled	—	0.035—0.05%	—
Oxydemeton-methyl	—	0.75—1.5 kg	—
Phorate	gran.	1—2 kg	—
Phosmet	follow local recommendations		
Phosphamidon	—	0.2—0.3 kg	—
SD 8280	—	0.8—1.0 kg	Broadcast at planting or prior to flooding
Triazophos	e.c.	0.75 kg	—

[1]DAT—Days After Transplanting

*These materials were used successfully under experimental conditions, it is important to transplant within 3 days of application or some phytotoxic effects may be found.

N. nigropictus

Vector of dwarf, transitory yellowing, tungro, yellow dwarf and yellow-orange leaf (Fig. 48).

Distribution

Australia, Bangladesh, Burma, Cambodia, Caroline Is., China, India, Indonesia, Laos, Malaysia, Mariana Is., Nepal, Nicobar Is., Pakistan, Papua, New Guinea, Philippines, Singapore, Sri Lanka, Taiwan, Vietnam, West Irian (CIE Map 286).

Control

Cultural

Clear weeds such as; *Panicum, Cyperus, Poa* which are alternative hosts.

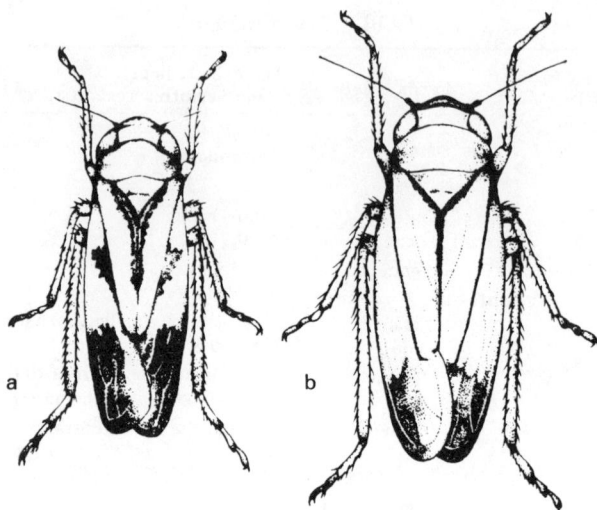

Fig. 48. Green rice leafhopper, *Nephotettix nigropictus,* (a) male adult, (b) female adult. (IRRI)

Chemical

See Table 5.

N. parvus

Vector of tungro and yellow dwarf virus diseases.

Distribution

India, Malay Peninsular, Sri Lanka.

Control

Control measures for this species have not been reported.

N. virescens

Vector of tungro, leaf yellowing, transitory yellowing, yellow dwarf and yellow-orange leaf virus diseases.

Distribution

Bangladesh, Burma, Cambodia, China, India, Indonesia, Laos, Malaysia, Pakistan, Philippines, Taiwan, Thailand, Vietnam (CIE Map 287).

Control

Cultural

Keep the area free from grassy weeds. Use rice varieties resistant to tungro and to the leafhopper.

Chemical

See Table 5.

Recilia dorsalis (Hemiptera: Cicadellidae)
Zig-zagged leafhopper

Vector of dwarf, orange leaf, tungro and yellow-orange leaf virus diseases.

Distribution

India, Japan, Malaysia, Philippines, Sri Lanka, Taiwan, Thailand.

Life cycle

The forewings of the adult insect (Fig. 49) are white, with light brown bands taking the shape of a 'W' and giving the wing a zig-zagged pattern. Body length is 3.5—4.0 mm (Plate 2).

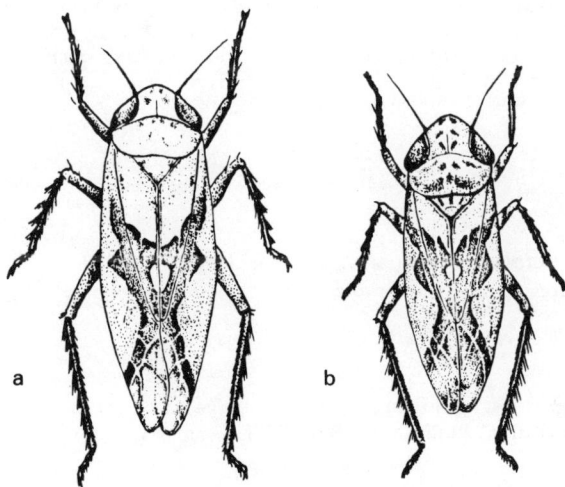

Fig. 49. Zig-zagged leafhopper, *Recilia dorsalis*, (a) female adult, (b) male adult. (IRRI)

Eggs are laid in rows within the leaf sheaths and hatch in 7—9 days. There are five nymphal stages lasting 16—18 days before the adult stage. The nymphs are yellowish brown in colour, varying from 1.0—3.0 mm in body length.

Control

Chemical

The insecticides recommended for the control of *Nephotettix* spp. should control *R. dorsalis*.

Unkanodes albifascia (Hemiptera: Delphacidae)

Vector of black-streaked dwarf and stripe virus diseases.

Distribution

Japan.

Symptoms

The viruses, once acquired, can be transmitted for the rest of the insects life. Stripe virus can be passed via the egg to the next generation. See p. 85 for a description of the disease symptoms.

Life cycle

In field populations only brachypterous (short-winged) forms have been found but in high density laboratory populations adults with full size wings were reared. At 25°C the egg and nymphal stages lasted 14 and 19 days respectively. Most adults live for 5 weeks, the females can lay 100—200 eggs each. Fourth instar nymphs overwinter in diapause, the adults emerge early in April.

Control

See recommendations for other planthoppers.

Laodelphax striatellus (Hemiptera: Delphacidae)

Small brown planthopper

Vector of black-streaked dwarf and stripe virus diseases.

Distribution

Europe, Japan, Korea, Philippines, Siberia, Taiwan (CIE Map 201).

Life cycle

This species has a varied range of host plants including rice, sugarcane, wheat, *Alopecurus* spp. and *Eragrostis* spp. and is found in a wide range of habitats.

The adult males (Fig. 50) are macropterous or brachypterous and usually 3.5 mm long, while the female is smaller, about 2.0 mm long. Sometimes individuals have red eyes. This species hibernates as diapausing fourth instar nymphs. The first adults move onto the transplanted rice crop from wheat. Usually 5% of these first adults carry stripe virus.

Eggs are laid in masses on the leaf. The nymphs are usually smaller than those of other species at corresponding instars.

By trapping these insects with yellow pans very early in the season the migration to rice can be forecast. The numbers trapped reach a peak just before mass migration. In Japan forecasts of rice stripe virus attack are made 6 months ahead and depend upon the number of nymphal *L. striatellus* present in the field in November and the percentage of them that are viruliferous.

Control

Chemical

These insects are important because of their virus transmitting ability. Sprays should be applied before the appearance of first generation adults. See Table 6 for recommended pesticides.

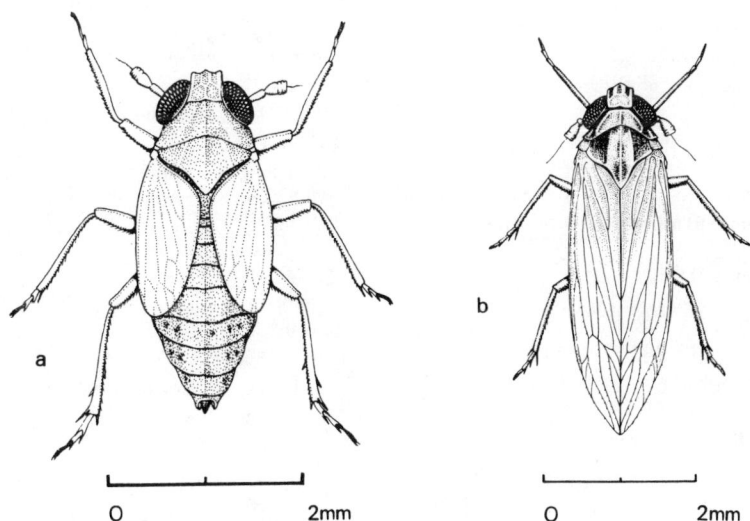

Fig. 50. Small brown planthopper, *Laodelphax striatellus*, (a) nymph, (b) adult. (H.R.B.)

TABLE 6. RECOMMENDATIONS FOR CHEMICAL CONTROL OF PLANTHOPPERS*

Insecticide	Formulation	Active ingredient per hectare	Application
Acephate	dust, gran. w.p.	0.3—0.5 kg	—
Allyxycarb	manufacturers recommendations		
Azinphos-methyl	—	0.3—0.5 kg	—
BHC	gran.	1.2—1.8 kg	Apply to soil surface at puddling when transplanting
BHC + isoprocarb	gran.	16—24 kg (formulation)	When stem borers are also a problem
BHC + MTMC	gran.	30 kg (formulation)	10—15 DAT[1] or 30—35 days after sowing
BPMC	fine gran.	0.9—1.2 kg	—
Bufencarb	e.c.	0.75 kg	—
Carbaryl	dust, w.p.	follow manufacturers recommendations	
Carbofuran	gran.	0.5 kg	At 15, 45 and 75 DAT on paddy and with seed at sowing followed by a second application 30 days after emergence on upland rice
Carbophenothion	e.c.	0.75 kg	—
Chlordimeform	—	0.75 kg	—
Chlorfenvinphos	gran. e.c.	2.4 kg 0.5—1 kg	Be careful to evenly distribute higher rates
Demeton-S-methyl	—	0.75—1.5 kg	—
Diazinon	gran. fine gran.	1.2—1.8 kg 0.9 kg	Apply to soil at puddling when transplanting
Diazinon + BPMC	fine gran.	30—40 kg (formulation)	—
Diazinon + MTMC	fine gran.	30—40 kg (formulation)	—
Dichlorvos	—	0.3—0.5 kg	—
Dicrotophos	—	0.29—0.58 kg	—
Disulfoton	gran.	1—1.5 kg	Broadcast at planting or prior to flooding
Endosulfan	e.c.	0.75 kg	—
Ethyl DD	e.c.	0.75 kg	—
Fenitrothion + BPMC	fine gran.	30—40 kg (formulation)	—
Fenitrothion + malathion	e.c., u.l.v.	1—2 l (formulation)	—

TABLE 6—continued

Insecticide	Formulation	Active ingredient per hectare	Application
Fenitrothion + MPMC	fine gran.	30—40 kg (formulation)	—
Fenthion	fine gran.	1—1.5 kg	—
Isoprocarb	w.p.	0.75 kg	—
	gran., dust	0.6—0.8 kg	—
Isothioate	gran.	3 kg	At 3 leaf stage
Malathion	u.l.v.	0.54 *l*	
Monocrotophos	e.c.	0.75 kg	—
	gran.	1—1.5 kg	—
MPMC	fine gran., gran.	0.6—0.8 kg	—
MTMC	fine gran., gran.	0.9—1.2 kg	—
MTMC + chlordimeform	fine gran.	30—40 kg (formulation)	—
Naled	manufacturers recommendations		
Oxydemeton-methyl	—	0.75—1.5 kg	—
Phorate	gran.	1—2 kg	—
Propaphos	dust	0.6—1 kg	—
Propoxur	w.p.	1—2 kg	—
Phosmet	follow local recommendations		—
XMC	—	1—2 kg	—

*When applying against brown planthopper be sure to treat the base of the plants.
[1]DAT—Days After Transplanting

Sogatella furcifera (Hemiptera: Delphacidae)
White-backed planthopper

Distribution

China, Fiji, India, Indonesia, Japan, Korea, Malaysia, Philippines, Sri Lanka, Taiwan, USSR, Vietnam (CIE Map 200).

Symptoms

Yellowing of the lower leaves, presence of sooty mould, followed by withering. The crop does not grow as fast as it should, the number of tillers is reduced by early infestation and the number of panicles by later attacks. The yield and quality of grain are reduced.

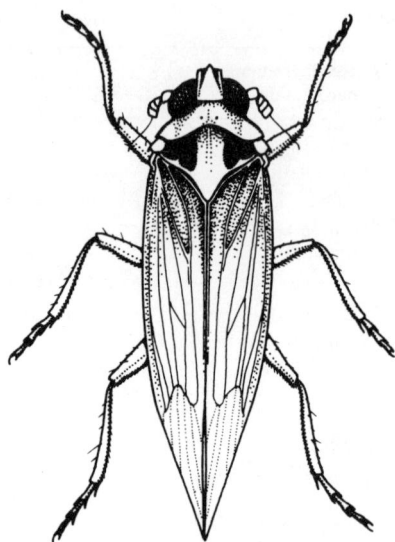

Fig. 51. White-backed planthopper,
Sogatella furcifera, adult. (H.R.B.)

O   2mm

Life cycle

This species is generally found during the early growth stages of the rice plant and multiplies very rapidly when rice is in standing water because of the high humidity. It has not been recorded as a vector of virus diseases.

The adult (Fig. 51) has a distinctive rather long narrow face. The forewings are uniformly hyaline with dark veins. There is a conspicuous black dot about the middle of the posterior edge of each forewing which meets when the forewings come together. The pronotum is pale yellow, the body is black dorsally and creamy white elsewhere.

The eggs, which hatch in 6 days, are laid in masses on the leaf sheath. Nymphs are pale to light brown in colour and range in size from 0.6 mm when young and newly hatched, to 2.0 mm after 11–12 days.

Control

Chemical

See Table 6.

142

Nilaparvata lugens (Hemiptera: Delphacidae)

Brown planthopper

Vector of grassy stunt virus disease.

Distribution

China, Fiji, India, Indonesia, Korea, Malaysia, New Guinea, Philippines, Sri Lanka, Taiwan, Thailand (CIE Map 199).

Symptoms

Plants develop 'hopperburn' (Fig. 52) due to leaves drying and browning after this insect has fed from them. Brown planthoppers can be found at the base of the plants, sooty moulds develop on the insects' excreta. The plant is most sensitive to attack at 26–39 days old. Typically patches of lodged 'burned' plants can be seen (Plate 2).

Life cycle

The adult (Fig. 53) is either light or dark brown in colour, and measures about 3 mm in body length.

The eggs are whitish or transparent, and are laid inside the leaf sheath and on the midrib of the leaf blade, in batches with their anterior ends attached to one another, they hatch in 8–9 days. Nymphs are white, 0.6 mm in body length when newly hatched, later become brown, and grow to 30 mm by the fifth and final moult. The nymphal stages last 12–13 days. In Japan the main source of infestation for the young rice is adults migrating from China, Indonesia, Vietnam and the Philippines. Losses are kept down by applying insecticides when the forecasting system, based on catches in light traps and tow nets, indicates population peaks, which usually occur at transplanting and two months later.

Control

Cultural

Where possible plant brown planthopper resistant varieties of rice. Destroy as far as possible all sources of the virus and alternative hosts by ploughing in stubble to bury infected plants and prevent growth of ratoon rice.

The fungus *Entomophora coronata* has shown promise as a biocontrol agent against *N. lugens* and further investigations into its use may be made.

Chemical

A 24 h seedling root soak of 90 g a.i./ha carbofuran in water to cover roots of the bundled seedlings has been recommended as an alternative to early treatment of the transplanted rice.

The use of insecticides to control this insect may not be successful or economic. If a susceptible rice variety is used and an *epidemic* of brown planthopper occurs

Fig. 52. Hopper burn — the browning, yellowing and drying of rice plants as a result of severe planthopper attack. (L.T. Kok)

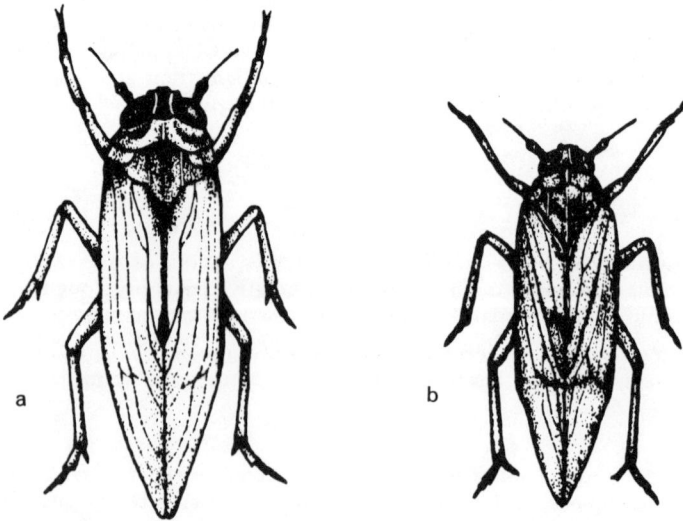

a   b

Fig. 53. Brown planthopper, *Nilaparvata lugens,* (a) female adult, (b) male adult. (IRRI)

early in the season it is possibly cheaper to abandon the first crop and replant with a resistant variety. Twenty or more brown planthoppers per hill indicate that treatment of some kind should be carried out. See Table 6 for recommended chemicals, treatment is more effective if applied at peak nymphal density i.e. when third instar nymphs predominate.

Sogatodes orizicola (Hemiptera: Delphacidae)
Sogatodes cubanus
Rice delphacid

Vectors of hoja blanca virus disease.

Distribution

South USA, Mexico, Central America, Argentina, Brazil, Colombia, Surinam, Caribbean (CIE Map 202).

Central America, Caribbean Is., W. Africa (CIE Map 224).

Life cycle

S. cubanus can feed on rice but usually infests grasses e.g. *Echinochloa*, which is a host of hoja blanca. The virus is transmitted to the rice crop by *S. orizicola* which can feed on both plants but prefers rice. The females lay 300—350 eggs in batches of 7 on the midribs of rice leaves. The eggs are 0.1 mm long, cylindrical, slightly curved and white when laid but darken with age.

Both *S. orizicola* and *S. cubanus* can diapause in the egg stage, but nymphs usually hatch in 4—8 days. The populations increase as the crop matures.

S. orizicola adults are 3—4 mm long, basically brownish red in colour with a light median stripe dorsally. The abdomen darkens towards the anus. The face has a light median stripe which leads to a white area at the top of the head. The wings are yellowish, the veins yellow except for some in the apical area and the white costal veins that form the dorsal stripe when the wings are folded over the body.

S. cubanus is usually smaller than *orizicola*, about 2 mm long, it is generally brown in colour. When the wings are folded dark spots on them form a saddle-shaped stigmata.

Both species are rather sedentary in habit, their spread is helped by heavy winds and flowing water.

Control

Cultural

Remove alternative hosts of hoja blanca i.e. grass weeds. Avoid successive plantings. Some rice varieties are resistant to the virus disease and should be used if hoja blanca is a problem.

TABLE 7. RECOMMENDATIONS FOR CHEMICAL CONTROL OF *SOGATODES SPP.*

Insecticide	Formulation	Active ingredient per hectare	Application
Allyxycarb	—	0.3—0.5%	Follow the manufacturers recommendations
BHC + MTMC	—	30 kg (formulation)	10—15 DAT[1] Flood the field to 8 cm and retain the water for 3—4 days after application
Chlorpyrifos	—	0.1—0.5 kg	—
Demeton-S-methyl	—	0.75—1.5 kg	—
Dichlorvos	—	0.3—0.5 kg	—
Dicrotophos	—	0.2—0.3 kg	—
Disulfoton	gran.	1—1.5 kg	Broadcast at planting or prior to flooding
Fenitrothion + malathion	u.l.v.	1—1.25 *l* (formulation)	When the infestation appears and repeat if necessary
Malathion	e.c.	0.5—0.75 *l*	—
	dust	0.32—0.4 kg	—
Oxydemeton-methyl	—	0.75—1 kg	—

[1]DAT—Days After Transplanting

Chemical

See Table 7.

Rhopalosiphum padi (Hemiptera: Aphidae)

Distribution

Africa, America, Asia, Australia, Europe (CIE Map 288).

Symptoms

Recent studies in Europe have shown that *R. padi* transmits the disease giallume (see p. 81).

Control

Cultural

Aphelinus sp. sometimes reduce the *R. padi* population considerably.

Chemical

In Europe demeton-methyl is used to control this aphid.

Hemiptera: Heteroptera

Several species of plant bugs feed on the sap of the rice plant, causing weakening of the plants and poor grain formation.

Leptocorisa acuta   **Leptocorisa lepida**
Leptocorisa costalis   **Leptocorisa tagalica**
Leptocorisa discoidalis   **Leptocorisa chinensis**
Leptocorisa oratorius   **Stenocoris southwoodi**   (Hemiptera: Alydidae)

Rice bugs

Distribution

Leptocorisa acuta (Fig. 54a) is a major pest of rice and found in all rice growing regions of the Far East; India, Sarawak, New Guinea, Samoa, New Caledonia and Fiji. *L. oratorius* is also widely distributed in the rice growing areas of the world, its range extends to Sri Lanka in the west, Malabar, Pakistan, Bhutan and Komai (Tibet) in the north, south to North Queensland and east to the Solomon Islands. *Stenocoris southwoodi* (Fig. 54b) is important in the African equatorial area. *Leptocorisa lepida* is common in India and found in Malaya and Burma and as far to the east as Thailand, *L. costalis* is recorded from Sumatra, Thailand and Malaya in the west to Borneo and the Philippine Is. *Leptocorisa tagalica* is found from the Philippine Is. in the north to the Halmahesa Is. in Indonesia in the south. *Leptocorisa discoidalis* has been found from Indonesia east to New Hebrides. *Leptocorisa chinensis* is established in Bhutan and China to the north, as far south as Sumatra and as far east as Malaya (CIE Map 225, *L. acuta*).

Symptoms

These insects feed on developing grain, during the day especially in the early morning and at dusk or when the sunlight is not intense. Grain that is soft and milky is vulnerable to attack. During the hottest part of the day the insects can be found sheltering on the underside of leaves. As a result of their activities panicles may bear partially or entirely empty grains. Brown spots occur where the insects have fed. In addition the grain may smell unpleasant, this smell can be passed on if grain is bulked and may lower the market value. The presence of the bugs in the field is indicated by this smell.

147

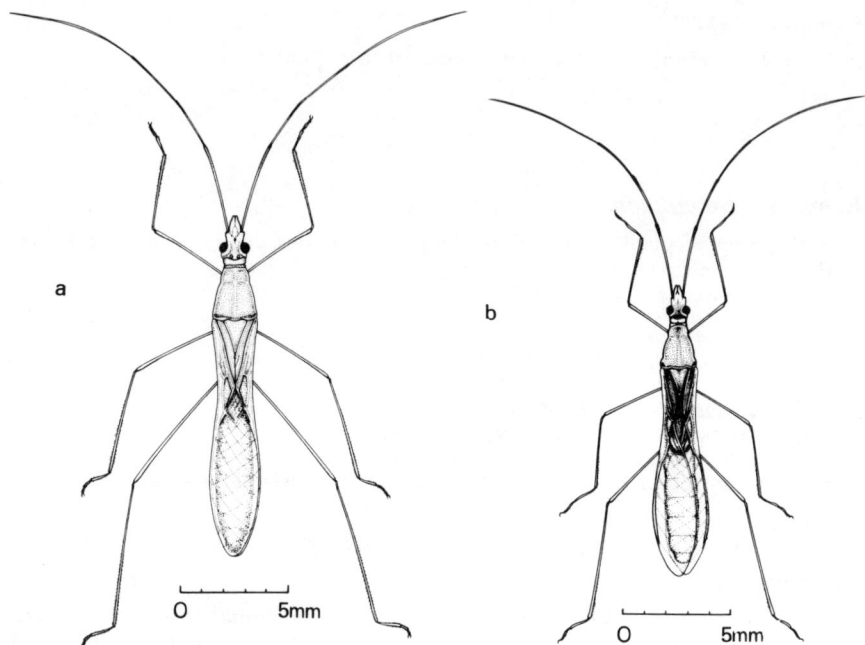

Fig. 54. Rice bugs (a) *Leptocorisa acuta,* adult, (b) *Stenocoris southwoodi.* (H.R.B.)

Life cycle

Adult rice bugs are about 15 mm long, slender and brownish-green in colour.
These bugs can survive on weeds in the absence of rice plants. They can live
up to 115 days when environmental conditions are favourable.

Eggs are deposited in rows on the leaves. They are somewhat flattened and
creamish-white when newly laid but become darker near hatching. Incubation
takes 5—8 days. There are five nymphal instars and distinct colour changes
follow each of the four moults. A newly hatched nymph is green but becomes
brownish as it grows. The nymphal period usually lasts for 17—27 days.

Control

Chemical

If insecticides are applied the application should be made at the flowering stage,
preferably in the evening. See Table 8 for the chemicals used to control this
group of insects.

TABLE 8. RECOMMENDATIONS FOR CHEMICAL CONTROL OF RICE BUGS

Insecticide	Formulation	Active ingredient per hectare	Application
Acephate	dust	follow manufacturers recommendation	
Azinphos-methyl	dust	0.3—0.5 kg	—
Chlorfenvinphos	e.c.	1—1.5 kg	Spray before peak nymph emergence
Dichlorvos	—	0.3—0.5 kg	Up to 2 applications depending on the infestation
Dicrotophos	—	0.25—0.5 kg	—
Fenthion	dust	0.6—1.2 kg	—
Malathion	follow manufacturers recommendations		
Monocrotophos	—	0.5 kg	—
MTMC	w.p.	0.75 kg	—
Phenthoate	—	0.5 kg	—
SD 8280	—	0.8—1.0 kg	—

Cletus trigonus (Hemiptera: Coreidae)
Slender rice bug, paddy ears

Distribution

North Borneo, India, Japan, Malaysia, Philippines, Sri Lanka, Taiwan.

Symptoms

Bugs are seen feeding upon young rice grains, causing discoloured spots. Such grain is known as pecky. Two groups of plants, the Gramineae and the Amaranthaceae are normally hosts of this insect.

Life cycle

Eggs are deposited singly and the first nymph emerges within 7 days. After 20 days the nymph has passed through 5 moults and become an adult. The adult is about 8 mm long, brownish yellow in colour with conspicuous brown punctures on its body. The pronotum projects laterally and very slightly towards the front.

Control

Chemical

Acephate or propaphos dust have been found suitable for the control of *C. trigonus*.

Oebalus pugnax (Hemiptera: Pentatomidae)

Rice stink bug

Distribution

Rice growing areas of Cuba, Dominican Rep., USA.

Symptoms

Nymphs and adults feed on rice grain at the milky stage. When the soft dough stage is reached damaged grain, if it develops fully, becomes pecky or spotted. Peckiness is due to a fungus infection entering through the feeding puncture. Pecky rice does not mill satisfactorily and there is a reduction in yield and quality.

Life cycle

Over-wintering adults (Fig. 55) emerge in early spring and feed on developing grass seeds. Adults are straw coloured, 9.5—12.5 mm long and about 5—6 mm broad. They have typical sharp shoulder spines which project forward and they give off a characteristic strong odour when disturbed. Eggs are 0.86 mm long and 0.65 mm in diameter and are laid in masses of 10—47. The eggs may be deposited on stems, leaves or panicles of rice or on several grass species and hatch in 4—8 days. They are bright green when laid but turn red before they hatch.

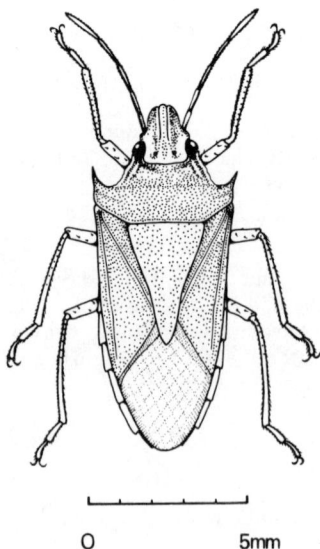

O   5mm

Fig. 55. Rice stink bug, *Oebalus pugnax,* adult. (H.R.B.)

There are 5 nymphal instars and the entire nymphal stage lasts 16—20 days. When the nymphs first hatch they have black thorax, head, legs and antennae and two black spots on their red abdomen. This colouration fades with successive nymphal stages. Adults live from 30—40 days but they can hibernate in wood trash and 'bunch' grass.

Control

Chemical

Carbaryl at 1.1 kg a.i./ha or chlorpyrifos at 0.8—1 kg a.i./ha should control these insects.

Oebalus ornata (Hemiptera: Pentatomidae)
Rice stink bug

Distribution

Brazil, Haiti, Dominican Rep., Puerto Rico.

Symptoms

Similar to those of *O. pugnax*.

Life cycle

Similar to *O. pugnax*, there may be 7 generations per year.

Control

See *O. pugnax*.

Oebalus poecilus (Hemiptera: Pentatomidae)
Distribution

Important throughout the South American rice growing area.

Symptoms

Symptoms are similar to those of *O. pugnax*. Milling quality is reduced in proportion to the number of bugs present at the milky stage of the rice grain.

Life cycle

The bugs are brown with yellow markings on the scutellum. Ten to 200 barrel-shaped eggs are laid in rows on the leaf and ears. After about 5 days nymphs emerge and begin to develop. Nymphal development lasts 15—20 days.

Control

See *O. pugnax*.

Scotinophara coarctata (Hemiptera: Pentatomidae)
Black paddy bug, the Malayan black rice bug

Distribution

India, Indonesia, Malaysia.

Symptoms

Nymphs and adults of this bug feed chiefly at the base of stems and often just below water level. Infected plants are often stunted. The leaves turn reddish-brown and grain does not develop. Direct injury to panicles is also common. Severe infestations may cause plant death. The bug is able to hide in cracks in soil during times of water stress but it appears to require a certain degree of moisture for its development.

Life cycle

The adult bug (Fig. 56) whose body length is 8–9 mm often appears in vast swarms and is strongly attracted to light.

It is brownish black with a few indistinct yellowish spots on the thorax, which bears spines below the anterior angles. Tibiae and tarsi are pinkish.

Eggs are deposited in batches of 40–50, each egg being 1 mm long and greenish or pinkish in colour. The incubation period is 4–7 days. The young nymphs are brown, with yellowish-green abdomen and some black spots. They moult four to five times and reach the adult stage in 25–30 days. Adults can live for up to 7 months.

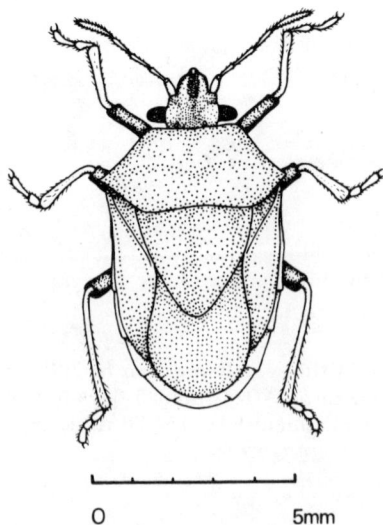

```
|——·——·——·——·——|
0              5mm
```

Fig. 56. Black paddy bug, *Scotinophara coarctata,* adult. (H.R.B.)

Scotinophara lurida (Hemiptera: Pentatomidae)
Japanese black rice bug

Distribution
India to Japan through all Asia.

Symptoms
Black bugs feeding at the base of the plant cause the ears and leaves to become chlorotic or reddish brown, heavily attacked leaves become twisted and yellow and dead hearts develop. If they feed very close to the growing point of the plant then tillering is reduced. These bugs have a distinctive smell. Injured grain is spotted with brown.

Life cycle
White eggs are laid in parallel rows on the leaves. Brown nymphs hatch after about 6 days. Nymphs moult 4 times becoming adults in 6—7 weeks. Only one generation per year occurs. Adults aestivate in the soil in the dry season.

Control of *Scotinophara* spp.
Cultural
These bugs tend to congregate at the base of the plants. In China *S. lurida* is recorded as laying 75% of its eggs within 10 cm of ground level. Recommended control measures are that the field should be flooded to 15 cm and allowed to stand for 1 day. This treatment is carried out 5 times, in the month of July, in China.

Chemical
In Japan, acephate, cyanofenphos, diazinon, dimethoate, EPN, fenthion, fenitrothion, malathion, mecarbam, phenthoate, trichlorphon and vamidothion have been recommended to control this pest. The manufacturers or local recommendations should be followed.

Diploxys fallax (Hemiptera: Pentatomidae)
Rice shield bug

Distribution
Malagasy, Swaziland.

Symptoms
Adults (Fig. 57) and nymphs feed on newly emerged florets before the grain reaches the milky stage, causing grain loss or deformation. A brown necrotic spot on the lemma is indicative of old feeding punctures.

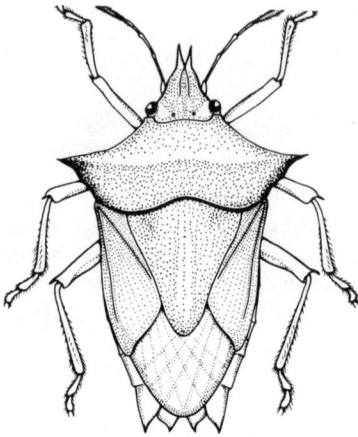

Fig. 57. Rice shield, *Diploxys fallax,* adult. (H.R.B.)

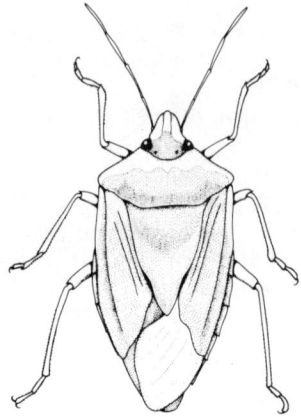

Fig. 58. Green vegetable bug, *Nezara viridula,* adult. (H.R.B.)

Life cycle

Before oviposition on rice adults feed on grass flowers. The eggs, about 20 in number, are laid in two rows on the upper leaf surface along the midrib. Nymphs are green in colour and adults light brown with a pair of lateral spines extending from the dorsal thorax.

Control

Cultural

Eradication of grasses on bunds and road verges would prevent the population of bugs from building up before the rice begins to flower.

Nezara viridula (Hemiptera: Pentatomidae)

Green vegetable bug, southern green stink bug

Distribution

This insect is found wherever rice is grown but does not always attack rice (CIE Map 27).

Symptoms

The insects (Fig. 58) can be seen in large numbers on the crop. At harvest some grain may be misshapen or speckled.

Life cycle

Hibernating adults emerge and feed on grasses and orchards before migrating to rice when about to breed. The pest is becoming increasingly serious in southern Japan where early planting of rice has provided a preferred host early in the season so that large populations of bugs are able to build up on later crops. Eggs are laid in parallel rows on the lower surface of rice leaves in masses of 20—130. The eggs are yellow when laid but turn red just before hatching. The nymphal period lasts for 35—45 days and during this time nymphs undergo 5 moults. Adult bugs exhibit four distinct colour phenotypes. They are 13—17 mm long and typically pentatomid. Females live about 30 days and lay their eggs toward the end of their lives.

Control

See Table 9.

TABLE 9. RECOMMENDATIONS FOR CHEMICAL
CONTROL OF *NEZARA VIRIDULA*

Insecticide	Formulation	Active ingredient per hectare	Application
Acephate	—	0.4—0.5 kg	When the insect is seen in large numbers
Azinphos-methyl	—	0.3—0.5 kg	—
Dichlorvos	—	1 kg	On stubble against over-wintering adults
Fenthion	dust	0.6—1 kg	—
Monocrotophos	—	0.5 kg	—
SD 8280	—	0.8—1.0 kg	—

Eusarcoris inconspicuus (Hemiptera: Pentatomidae)

La chinche del arroz

Distribution

Spain.

Symptoms

This pentatomid feeds on young grains, sucking out the soft interior. In some cases grain can recover from this attack but the surface becomes blotched with yellow, grey or blackish irregular spots due to fungus or bacterial infection at the point where the insect fed. In addition to this damage grain can be irregular in shape and may not form or ripen properly. The spotting reduces grain quality.

Life cycle

Eggs are laid on the upper surface of young leaves and 5–7 days later the first instar nymphs emerge. There are 5 immature stages. Adults are brownish ventrally on the abdomen and they have a white dentate lateral stripe.

Control

Chemical

1 *l*/ha u.l.v. phenthoate or 1.15 *l* a.i./ha fenitrothion have been used with success to control this insect.

Thysanoptera

Baliothrips biformis (Thysanoptera: Thripidae)

Rice thrips, paddy thrips

Distribution

Bangladesh, India, Indonesia, Japan, Malaysia, Sri Lanka, Thailand, Vietnam (CIE Map 215).

Symptoms

The leaves of young plants have fine yellowish or silvery streaks which later join together to colour the whole leaf. Later the leaves roll longitudinally from the margin towards the midrib and the whole plant withers. Leaves which are unopened when attacked usually remain so. Plants which can survive infestation are usually able to reach maturity and bear grains. This insect is essentially a pest of young (usually less than four weeks old) plants and is important because of its high rate of multiplication.

Life cycle

Eggs are inserted singly in leaf tissues and hatch in 3 days. Nymphs are whitish or pale yellow in colour and remain in the young rolled leaves where they

156

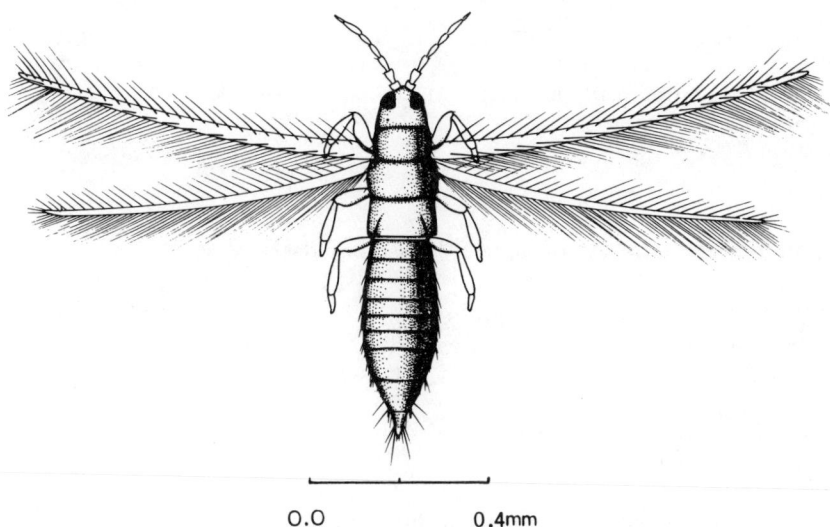

```
0.0              0.4mm
```

Fig. 59. Rice thrips, *Baliothrips biformis*, adult. (H.R.B.)

develop into adults. Usually four nymphal instars take place, the one before the adult being a resting stage. The nymphal stage lasts for 10—14 days and the average life-cycle is generally 2 weeks, although adults can live up to 3 weeks. The adults (Fig. 59) are minute elongate insects about 1 mm in length, dark brown in colour and with seven-jointed antennae. At the base of the forewing is a light spot.

TABLE 10. RECOMMENDATIONS FOR CHEMICAL CONTROL OF *BALIOTHRIPS BIFORMIS*

Insecticide	Formulation	Active ingredient per hectare	Application
Azinphos-methyl	—	0.3—0.5 kg	—
Carbaryl	follow manufacturers recommendation		
Diazinon	—	0.5—1 kg	—
Fenitrothion + malathion	e.c., u.l.v.	0.9—1.25 *l* (formulation)	—
Fenthion	dust	0.6—1.0 kg	—
Malathion	follow manufacturers recommendation		
Phorate	gran.	1—2 kg	—

Control

Cultural

Remove and destroy badly infested leaves in nursery beds.

Chemical

See Table 10.

Haplothrips aculeatus (Thysanoptera: Phlaeothripidae)

Distribution

Palearctic region.

Symptoms

Numerous thrips feeding on the flowers.

Life cycle

The eggs are deposited on the surface of the host plant, usually on the flowering heads. Nymphs feed mainly on the flowers, although the adults are apparently partly predacious.

Control

Similar to rice thrips control.

Lepidoptera

Latoia bicolor (Lepidoptera: Limacodidae)

Slug caterpillar, nettle grub

Distribution

India, Indonesia.

Symptoms

Young leaves are eaten by larvae 20–30 mm long, fat and bright yellowish green with three dorsal bluish stripes alternating with 4 dorsal and lateral rows of spines arising in tufts from tubercles. If the larvae are handled they cause an irritating rash.

Life cycle

Flat scaly eggs are laid on leaves in clusters of 10–30, after 6 to 8 days the larvae emerge. The larval period lasts nearly 50 days (in India). The larvae spin dark brown oval cocoons on the leaves or stems, in these cocoons pupation takes place. After nearly 40 days the adult emerges, head, thorax and forewings are pea green, abdomen, hindwing and underside are yellowish red. Wing span is

30—38 mm. In India three generations have been observed between June and November.

Control

Chemical

Dimethoate at 700 ml a.i./ha or malathion at 200 ml a.i./ha have been recommended to control this insect.

Cnaphalocrocis medinalis (Lepidoptera: Pyralidae)
Rice leaf folder, rice leaf roller, grass leaf roller

Distribution

India, Indonesia, Korea, Malaysia, Pakistan, Philippines (CIE Map 212).

Symptoms

Larvae infest the leaves, rolling them longitudinally and together and live inside the rolled leaf. Leaves are sometimes not actually rolled, but the tips are fastened to the basal part giving the rolled appearance. With heavy infestations plants appear scorched and sickly. High infestation can lead to severe yield losses, late crops are particularly liable to infestation.

Life cycle

The adult moth (Fig. 60) is often seen among the plants during the day. It is a brownish-orange colour, small, about 8—10 mm in length, with a wing expanse of 12—20mm. There are several distinct dark wavy lines on the wings, the outer margins of which are characterised by a dark brown to grey band.

The eggs are laid in batches of 10—12 arranged linearly along the midrib on either surface of the leaf. They are flat, oval in shape, yellow in colour and hatch in 4 days. The larva lives within the fastened leaves for 15—25 days before pupating there. The pupa is slender, greenish-brown coloured. After 6 8 days the moth emerges. The life cycle is generally 23—35 days.

Control

Cultural

Keeping the area free of weeds prevents a build-up of this pest on alternative hosts (Gramineae). In small areas of rice clipping of infested leaves and destroying them may control the pest.

Chemical

See Table 11. Apply insecticides only when the infestation is heavy.

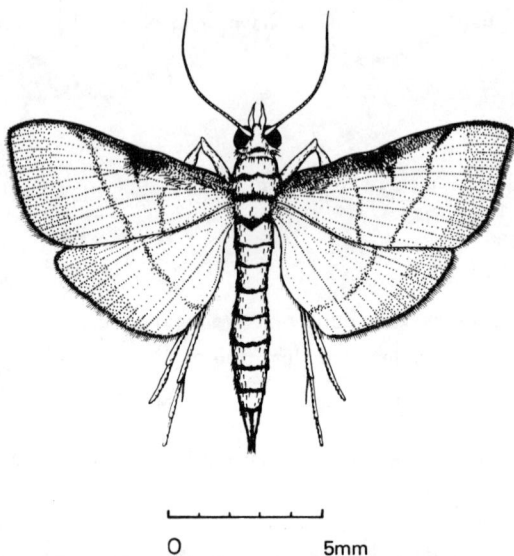

Fig. 60. Rice leaf folder or roller, *Cnaphalocrocis medinalis*. (H.R.B.)

TABLE 11. RECOMMENDATIONS FOR CHEMICAL
CONTROL OF *CNAPHALOCROCIS MEDINALIS*

Insecticide	Formulation	Active ingredient per hectare	Application
Azinphos-methyl	–	0.3–0.5 kg	–
Carbaryl	dust	1.25 kg	–
	w.p.	1.00 kg	–
Cartap	–	0.8 kg	–
Chlordimeform	–	0.5–0.75 kg	–
Dichlorvos	–	0.75–1 kg	Generally twice with a 7 day interval
Dimethoate	e.c.	0.5 kg	–
Formothion	follow manufacturers recommendations		–
Malathion	e.c.	0.5 kg	–
Monocrotophos	–	0.3–0.6 kg	–

Nymphula depunctalis (Lepidoptera: Pyralidae)

Rice caseworm

Distribution

Argentina, Australia, Brazil, Gambia, Ghana, India, Indonesia, Malagasy Republic, Malaysia, Malawi, Mauritius, Mozambique, Nigeria, Pakistan, Philippines, Sri Lanka, Uraguay, Venezuela, Zaire (CIE Map 176).

Symptoms

This is a pest of rice in irrigated areas. The larva cuts off tips of leaves to make tubes or cases in which it lives. Each larva attacks several plants and constructs many cases before it is fully grown. Damage is especially severe to young plants. Larvae feed on the leaves and the papery membrane on the upper epidermis that remains appears white (Fig. 61) at a distance. Attacked plants often become stunted (Plate 3).

Life cycle

The adult (Fig. 62) is nocturnal and attracted to light. It is a small, delicate insect, snowy white in colour, with pale brownish-yellow spots on both fore and hind wings, wing expanse is 15 mm. It can live for up to 3 weeks. Eggs are laid in one or two adjacent rows on the leaf sheath and they hatch in about 3 days. Larvae are transparent green in colour with light brownish-orange heads. They

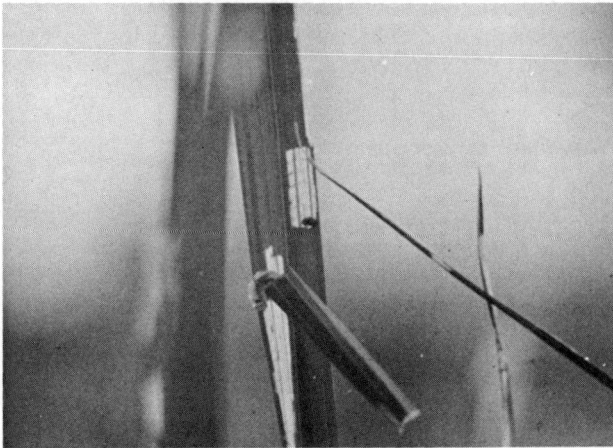

Fig. 61. Larva of *Nymphula depunctalis* eating the chlorophyll of a rice leaf and leaving characteristic feeding marks. (L.T. Kok)

161

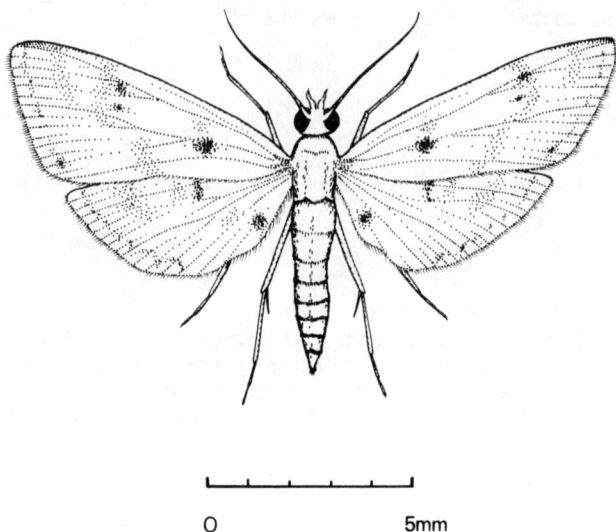

Fig. 62. Rice caseworm, *Nymphula depunctalis*. (H.R.B.)

are 20 mm long when fully grown, and are semi-aquatic in habit with lateral slender filamentous gills. The larval stage lasts for 15—30 days. Pupation takes place in a brownish tube of leaves which is attached to the basal part of the tillers. The adult emerges in 4—7 days, and the complete life cycle takes 19—37 days.

Nymphula nympheata (Lepidoptera: Pyralidae)
China mark moth

Distribution

Hungary, Italy.

Symptoms

Feeding damage and larvae in their cases will be seen.

Life cycle

Similar to *N. depunctalis*. Five generations per year are usual in Italy, the first is the most important because it attacks very young rice plants.

Nymphula vittalis (Lepidoptera: Pyralidae)
Smaller rice caseworm

Distribution

Japan.

Symptoms

The larvae fold the leaves of the plant.

Control of *Nymphula* **spp.**

Cultural

Since the larvae are aquatic draining the field for 3—5 days will kill them.

Chemical

In Japan fenthion or trichlorphon dusts or e.c. have been found to control this pest, the manufacturers recommendations should be followed regarding quantities. Dichlorvos at 0.5 kg a.i./ha is recommended by the manufacturers.

Parnara guttata (Lepidoptera: Hesperiidae)
Rice skipper

Distribution

Celebes, China, Himalayas, India, Indonesia, Japan.

Symptoms

The larvae feed on leaves from the margins inwards and then parallel to the midrib which is not usually eaten. In addition the larvae tie, with silk, two edges of the same leaf or two separate leaves together to form a tube in which they live. Sometimes tips of leaves are rolled backwards and tied to the broader portion to form the tube. Damage is particularly severe in young transplanted seedlings and an attack may last up to ear emergence. In cases of severe infestation, plants are sickly and often fail to recover. Grain quality is reduced and maturation is uneven.

Life cycle

Pale yellow eggs are deposited singly on the leaves and hatch within 3 days. The full grown larva (20—30 mm long) is pale green with a dark head and a T-shaped black spot on the rear end (anal flap). After about 30 days they pupate in the leaf tubes. Pupae are pale green to yellow in colour. Adults, which emerge after 8—10 days, are olive brown butterflies 10—15 mm long with a wing span of 20—40 mm. On the forewings are 5—9 small whitish spots and on the hind-wings are 2—4 similar specks.

Pelopidas mathias (Lepidoptera: Hesperiidae)
Rice skipper

Distribution

Burundi, China, Egypt, South India, Indonesia, New Guinea, Rwanda, West Africa.

Symptoms

Symptoms are similar to those of *Parnara guttata*.

Life cycle

Similar to that of *P. guttata*. The larvae are pale green with, on either side of the head, a vertical red streak between white lines. The adult differs from *P. guttata* in usually having only 2 specks in the discal cell and another 2 groups of 3 white specks, one anterior and one more marginal than the first group. The hindwings are olive brown, usually without specks.

Telicota augias (Lepidoptera: Hesperiidae)
Rice skipper

Distribution

Australia, Indian sub-continent, Java, Malaysia, Philippines.

Symptoms

Similar to those of *Parnara guttata*.

Life cycle

The adult butterfly (Fig. 63) is similar in size to the species discussed above, it has orange wings with brown markings. Fully grown larvae are 40 mm long, green in colour and with dark heads and with a black spot on the anal flap.

Control of rice skippers

Chemical

Azinphos-methyl at 0.3—0.5 kg a.i./ha; cartap at 0.8 kg a.i./ha; DDT at 0.15 kg a.i./ha; dichlorvos at 0.4—0.6 kg a.i./ha or pyridaphenthion + MTMC fine granules at the rate indicated by the manufacturer are all recorded as satisfactorily controlling this insect.

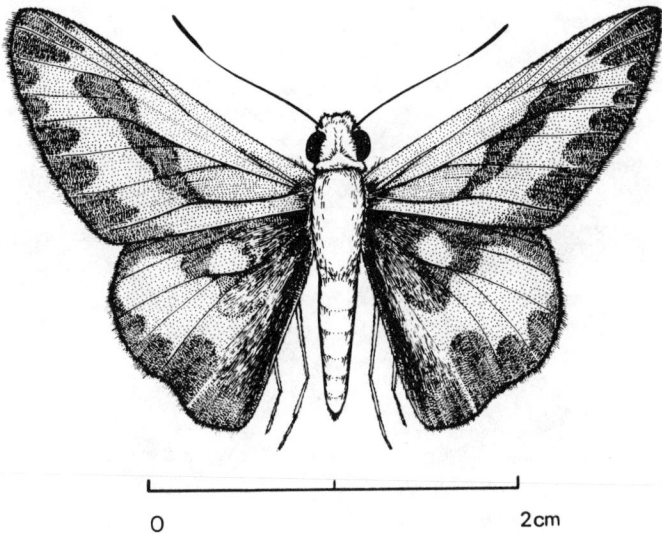

Fig. 63. Rice skipper, *Telicota augias*. (H.R.B.)

Spodoptera exempta (Lepidoptera: Noctuidae)
Spodoptera exigua
Armyworms

Distribution

Spodoptera exempta (Fig. 64) is found in Africa, Asia, Australasia and some Pacific islands (CIE Map 53).

S. exigua (Fig. 65) is found in Africa, North America, Asia, Australasia, Hawaiian Is., W. Irian, USSR, West Indies (CIE Map 302).

Symptoms

The damage is caused by gregarious larvae, moving in armies which completely eat young leaves and stems. Young seedlings which do not have well developed root systems are completely destroyed, older plants recover after attack and tiller vigorously. Upland rice is most commonly attacked.

Life cycle

The moths are dull brown insects, up to 25 mm wing-span, and have an active life of 8—10 days. Eggs are laid on the leaves of the host plant and hatch after 2—4 days incubation. Larvae are blackish and go through six stages, reaching a maximum length of 3.75—5.0 cm before pupation. The larval period lasts 10—12

165

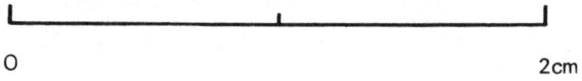

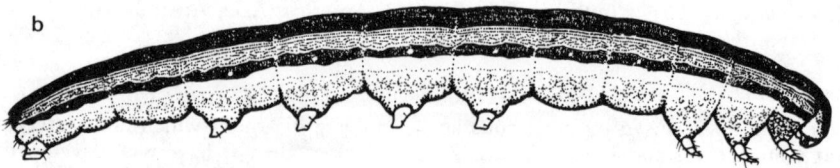

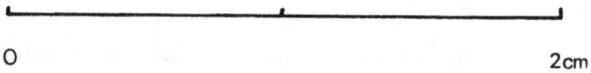

Fig. 64. Armyworm, *Spodoptera exempta*, (a) adult, (b) larva. (H.R.B.)

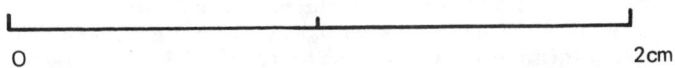

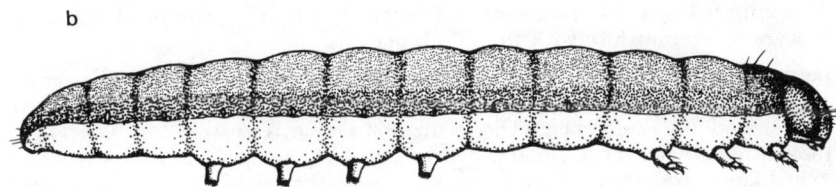

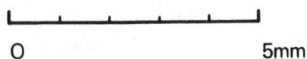

Fig. 65. Armyworm, *Spodoptera exigua*, (a) adult, (b) larva. (H.R.B.)

days, slightly longer with *S. exempta*. Pupation is in the soil and takes about 6 days, the total life cycle lasts about 3 weeks.

Natural predators such as ants and birds eat many of the swarming larvae, and as outbreaks always occur at the beginning of the rainy season when vegetation is lush, the normal onset of the heavy rains destroys large numbers of moths and larvae and after an initial outbreak there is rarely an important second generation.

Spodoptera mauritia (Lepidoptera: Noctuidae)
Rice armyworm, rice swarming caterpillar

Distribution

Africa, Australia, India, Indonesia, Malaysia, Pakistan, Philippines, Sri Lanka, Thailand, some Pacific islands (CIE Map 162).

Symptoms

Rice seedlings, either in seed beds or in fields are attacked by numerous larvae, feeding on the leaves, nibbling first at the surface and eating only small pieces. As they grow, they become voracious and can devastate a whole field within a short time and then move on to the next. They multiply very rapidly and spread easily, moving in large numbers, because of which they are often called swarming caterpillars. Feeding normally takes place after dark as the larvae are nocturnal in habit. Although mostly young seedlings are affected, older plants up to ear emergence stage have been attacked.

Life cycle

The adult is a greyish-brown moth (Fig. 66a), 25—30 mm long and 6 mm wide, with 35 mm wingspan. The forewings are marked with several wavy lines, in their centre is a distinct black spot, below which is a tiny white patch. The hind-wings are brownish-white with thin black margins.

Spherical eggs are laid in oblong clusters of 200—350 at the tip of the lower surface of an upright expanded leaf. They are covered by short buff hairs derived from the body of the female. The incubation period is usually 2—3 days; although some hatch in 5—9 days, depending on the temperature. Newly hatched larvae are green, about 2 mm long and rather difficult to detect against the leaves. As they grow their colouration becomes more varied, usually brownish or entirely brown. It takes 15—24 days for the larvae (Fig. 66b) to reach full size (35—40 mm long). The body is dark green dorsally with a pale line running down its back and greenish-yellow ventrally. Along each side of the body are three stripes, the second reddish, while the first and third are pale coloured. Just above the top stripe are very distinct semi-lunar or crescent shaped black spots. When fully fed the larva pupates in the soil. The pupa is dark brown, about 13 mm long and it has two slender spines at its narrow apex. The pupal stage lasts 7—14 days and the entire life cycle 37—40 days.

168

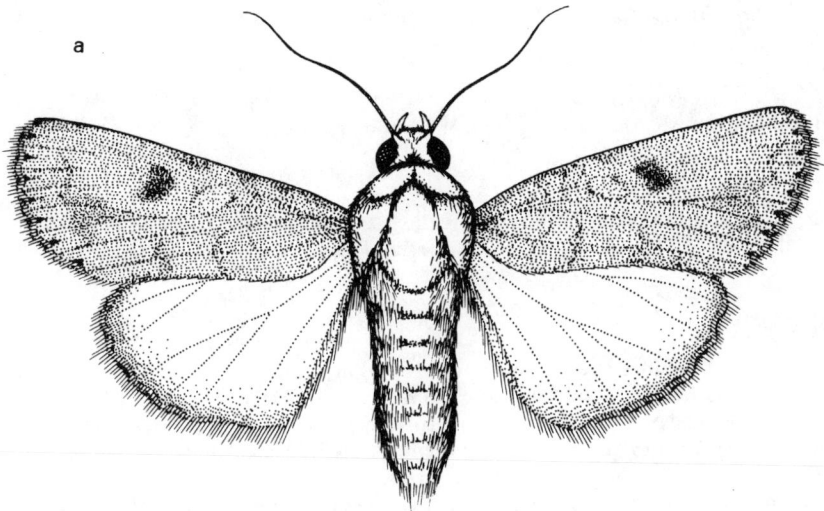

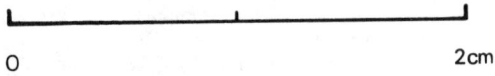

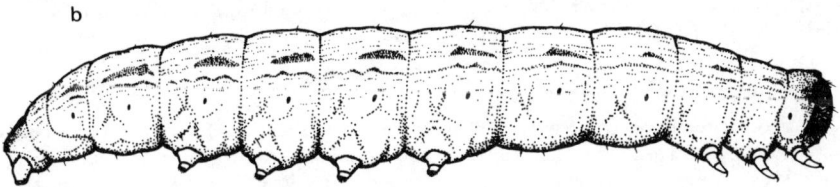

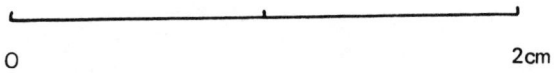

Fig. 66. Rice armyworm or swarming caterpillar, *Spodoptera mauritia*, (a) adult, (b) larva. (H.R.B.)

Spodoptera frugiperda (Lepidoptera: Noctuidae)
Fall armyworm
Distribution
Central and South America, Canada, West Indies, Mexico, USA (CIE Map 68).

Symptoms
This species is polyphagous, but may cause serious losses in rice late in the growing season. The larvae are non-selective feeders and many eat entire rice plants.

Life cycle
This species can hibernate as full grown larvae in the soil. It also pupates in small cells made in dry soil, and so cannot maintain successive generations in flooded rice fields. There is migration from other upland crops after pupation. Adult moths fly into the rice crop to deposit their eggs. Adults are ash grey in colour with irregular white and pale grey markings on the forewings. They are 2 cm long with a 4 cm wing span. Egg masses of 50 to several hundred eggs are laid on leaf blades and covered with scales from the female's body.

Newly hatched larvae are white with black heads. They are fully grown at 2–3 weeks by which time they are green to black with a white stripe down each side, below which is a band of yellow. Larvae may crawl up inside the rice leaf sheaths or move from plant to plant suspended by a silken thread. Large numbers of larvae may 'march' together to another food source. The number of generations each year depends on the climate. In the most favourable conditions there may be as many as six generations in one year.

Spodoptera litura (Lepidoptera: Noctuidae)
Rice cutworm, common cutworm
Distribution
Australia, India, Indonesia, Malaysia, Philippines, Thailand, some Pacific islands (CIE Map 61).

Symptoms
This species feeds on the leaves of a wide range of crops. The young larva feeds on leaf surfaces from the edge towards the mid-ribs. As it grows, it becomes voracious. Young seedlings are often cut at ground level while larger plants are defoliated.

Life cycle
The moth (Fig. 67a) has dark purplish-brown forewings with numerous spots and light coloured lines, which are more pronounced on the proximal half of the wing. The hindwings are whitish, narrowly banded along the outer margin. Its body length is 20–25 mm.

170

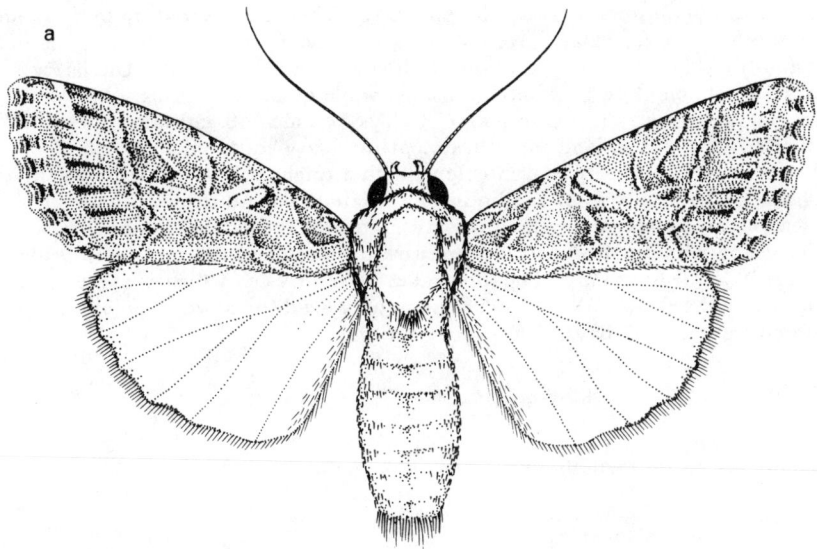

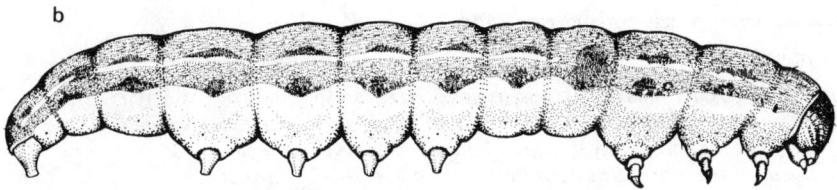

Fig. 67. Rice cutworm, *Spodoptera litura*, (a) adult, (b) larva. (H.R.B.)

Eggs are laid in clusters of 200—300 on leaves. They are covered by buff coloured hairs and hatch in 3—4 days. Newly-emerged larvae are small, blackish-green in colour with a distinct black band on the first abdominal segment. The larvae (Fig. 67b) remain close together for a short while before dispersing. Active at night, they become fully fed in about 20 days, reaching 40—50 mm in length. Mature larvae are stout and smooth, except for a few scattered short hairs, dull greyish and blackish-green in colour, with a bright yellow stripe down the back and along each side of the body. Each lateral yellow stripe is bordered by semi-lunar black spots. Along the lower edge of the side of the body is a duller yellow stripe. Head and legs are dark coloured and on the head capsule is a pale V-shaped marking. Pupation takes place in a cell which is made about 5 cm below the surface of the soil. The pupa is reddish-brown in colour and the moth emerges in 6—7 days. The whole life cycle takes about 30 days.

Control of *Spodoptera* **spp.**

Cultural

Flooding can save irrigated crops.

Chemical

Spray or dust with an insecticide as advised by local extension officers. Carbofuran, chlordimeform, dichlorvos, fenitrothion and several other insecticides have been used in various countries. Prompt application is important because the damage can rapidly increase.

Mythimna unipuncta (Lepidoptera: Noctuidae)

Common armyworm, ear-cutting caterpillar

Distribution

Southern Canada, USA, Hawaii, Central and some South American countries, Southern Europe, West Africa (CIE Map 231).

Mythimna separata (Lepidoptera: Noctuidae)

Rice armyworm, ear-cutting caterpillar

Distribution

Throughout Asia, Pacific Islands, east Australia, Fiji, New Zealand (CIE Map 230). In the past there has been a good deal of confusion over these species as they are very similar. Related species *Pseudaletia loreyi*, *P. venalba*, *P. irregularis*, *Prodenia eridania* and *Spodoptera pecten* have also been recorded damaging rice crops in various places. Their attacks may be sporadic but if heavy, an entire crop may be lost.

Symptoms

Mythimna spp. larvae are polyphagous and skeletonise leaves. Young larvae have only two pairs of prolegs and move with a characteristic looping action. Later larval stages become gregarious and voracious, eating whole leaves and plants usually at night. The sixth and final larval stage cut off rice panicles from the peduncle, hence the term 'ear-cutting'. It is this stage which causes the most serious losses to the rice crop.

Life cycle

The entire larval stage usually lasts 28 days. The larvae pupate in the soil. After 7—8 days moths emerge and begin egg-laying within three days. Males die after 3 days but females can live up to seven during which time they lay their eggs in batches of about 100 between the leaf sheaths and stems of upland rice or grasses. Later moths (there are 5 generations per year) or migrating larvae move into the rice crop.

Control

Cultural

Ploughing and burning stubble after harvesting, clean weeding.

Chemical

Dichlorvos at 0.6 kg a.i./ha, trichlorphon at 0.8 kg a.i./ha or fenitrothion at 1.15 *l* a.i./ha are recommended.

Naranga aenescens (Lepidoptera: Noctuidae)
Green rice caterpillar
Distribution

China, Japan, Vietnam.

Symptoms

Occasionally the green larvae of this moth become numerous and attack rice, mainly the seedlings of paddy rice.

Control

Chemical

See Table 12.

Diptera
Agromyza oryzae (Diptera: Agromyzidae)
Japanese rice leaf miner
Distribution

Northern Japan, Java, eastern Siberia.

TABLE 12. RECOMMENDATIONS FOR CHEMICAL CONTROL OF *NARANGA AENESCENS*

Insecticide	Formulation	Active ingredient per hectare	Application
Cartap	—	0.6 kg	—
Fenitrothion	e.c.	0.5 *l*	—
Fenitrothion + malathion	e.c.	1.5 kg (formulation)	—
	u.l.v.	1.5 *l* (formulation)	—
Phenthoate	follow manufacturers recommendations		

Symptoms

Leaves become withered and larvae can be observed mining in them. The mines are narrow and winding. This damage causes a reduction in shoots and a delay in ear development. Nursery beds of rice are particularly at risk.

Life cycle

White elliptical eggs (1 mm long) are laid singly near the tip of a leaf, larvae emerge after 6 days and begin to mine downwards. Larval development lasts 10—14 days. Pupation may take place within or on the surface of the leaf and lasts 7—18 days. There are 3 generations during a year.

Control

Chemical

This pest should be controlled when the rice is young, see Table 13.

TABLE 13. RECOMMENDATION FOR CHEMICAL CONTROL OF *AGROMYZA ORYZAE*

Insecticide	Formulation	Active ingredient per hectare	Application
Cartap	50% e.c.	0.4 kg	—
Dichlorvos	—	0.6—0.8 kg	—
EPN	—	0.3 kg	—
Fenthion	dust	0.6—1.2 kg	—
Phosmet	follow local recommendations		—
Trichlorphon	—	0.08 kg	—

Hydrellia griseola (Diptera: Ephydridae)
Rice leaf miner, smaller rice leaf miner, leaf-boring maggot

Distribution

North Africa, South America, Europe, Japan, Korea, Malaysia, USA.

Symptoms

Damage is caused by larvae which bore into the leaves and feed on the mesophyll tissues. The leaves subsequently shrivel and lie on the surface of the water. The larvae also may mine the leaf sheaths.

The early mine is 0.1—0.2 mm wide and usually linear, appearing as a whitish streak. Later it widens to a blotch as the larva moves about.

Damage is particularly severe if the attack occurs at the seedling stage. Surveys in Japan indicate that direct seeded rice may be more at risk than transplanted rice.

Life cycle

The number of generations per year varies with the location of the crop and the climate, e.g. 8 generations in northern Japan, 11 in California. An aquatic habitat is preferred. Flies (Fig. 68) can withstand submergence and are able to walk on the water surface. The adult is the overwintering stage and looks like a small housefly. Males, which are slightly smaller than the females are about 2 mm long and light grey in colour. There is a very characteristic frontal lunule, golden or shining white in colour. The wing span is 2.5—3.2 mm. Each female lays 50—100 eggs singly on the upper surface of the leaves close to the water surface. Humidity is very important for hatching, which is optimal at 98% r.h. and the period of incubation varies with the humidity of the environment.

Larvae are 0.10—0.17 mm wide and 0.33—1.33 mm long on hatching, and almost transparent to light cream in colour, they bore immediately into the leaf and pass all three larval instars there. This takes 7—10 days, or longer in cooler weather. Pupation takes place inside the leaf. The puparium measures from 3.10—4.25 mm in width, 3.61 mm in length. It is ovoid, tapering and light to dark golden brown in colour. The pupae can be seen inside the transparent mines. They have occasionally been found on the soil surface.

Adults begin egg laying 3—4 days after emergence, and can live for 3—4 months.

Control

Chemical

Routine treatments to control other insects such as stem borers will usually control *H. griseola*. Insecticides recommended for *H. griseola* control are given in Table 14.

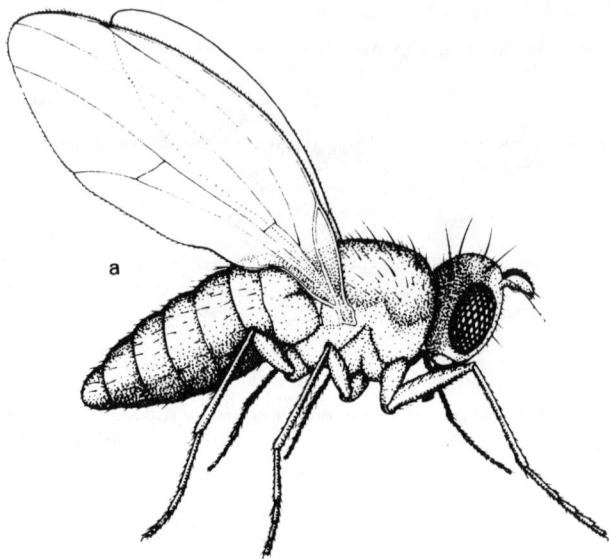

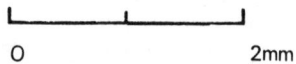

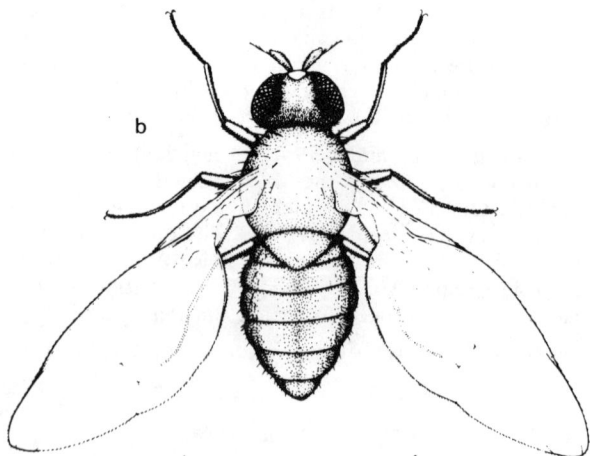

Fig. 68. Rice leaf miner, *Hydrellia griseola*. (H.R.B.)

TABLE 14. RECOMMENDATIONS FOR CHEMICAL CONTROL OF *HYDRELLIA GRISEOLA*

Insecticide	Formulation	Active ingredient per hectare	Application
Cartap	—	0.35 kg	—
Diazinon	gran.	2 kg	Broadcast evenly 75 DAT[1], retain water in the field for 3 days after treatment
Dichlorvos	—	0.6–0.8 kg	—
Fenitrothion	dust	0.5 kg	—
	e.c.	1 kg	—
Fenthion	dust	0.6–1.2 kg	—
Leptophos	follow manufacturers recommendations		
Phorate	gran.	1–2 kg	—
Trichlorphon	—	0.8–1 kg	—

[1]DAT — days after transplanting

Hydrellia philippina (Diptera: Ephydridae)
Rice whorl maggot

Distribution

Philippines.

Symptoms

Plants are stunted and do not tiller satisfactorily. Leaves are disfigured by whitish blotches. Newly planted seedlings are susceptible to attack but after 6 weeks this pest no longer causes damage.

Life cycle

Adults are dull grey flies about 2 mm long. The white cylindrical eggs are laid singly on either surface of the leaves and larvae emerge after 2–6 days. When they hatch the larvae migrate to the unopened central whorl of leaves where the entire larval period (10–12 days) is spent. Larvae feed on unopened leaves, nibbling the inner margins. When the leaves develop and open the blotches can be seen (Plate 3).

Control

Chemical

Granular insecticides such as diazinon or carbofuran applied shortly after transplanting.

Hydrellia sasakii (Diptera: Ephydridae)

Paddy stem maggot

Distribution

Japan.

Symptoms

This pest is particularly important on the late planted crop in Japan. It is most damaging early in the life of the crop, and is abundant in the fields for about one month after transplanting. The larvae feed on leaf blades and on the margins of unopened leaves. Leaves are marked with small spots and stripes. A heavy attack stunts the crop and reduces yield. The percentage yield reduction is 0.3 of the percentage of leaves having large spotted marks on the marginal part of the laminae.

Life cycle

Adults fly into the crop soon after transplanting. They can also live on grass species, e.g. *Leptochloa chinensis, Leersia sayanuka* and *L. japonica.* Eggs are laid singly on both sides of the leaf and hatch in about 2 days. The larval stage lasts 2—3 weeks and pupation occurs between the leaf sheath and stem. The pupal stage varies from 17 days in spring to 5—8 days later in the year. In southwest Japan the insect has 5 generations, the third and fourth doing the most damage to rice. In the winter the larval stage hibernates on *Leersia sayanuka* or other species mentioned above.

Control

Cultural

For oviposition the fly prefers rice planted with wide spacing, the farmer should therefore aim to get a well covered field as quickly as possible. The use of heavy tillering rice varieties, planting several seedlings per hill and good water and fertilizer management will help him to achieve this.

Chemical

When puddling the soil just before transplanting apply dimethoate granules at 1 kg a.i./ha.

Ephydra spp. (Diptera: Ephydridae)

Symptoms

Larval injury to young plants can lead to wilting and death.

Control

Cultural

In Egypt draining the field for 4 days, 10 days after transplanting was found to control an *Ephydra* sp.

Chemical

In Spain trichlorphon applied after germination controls the *Ephydra* sp. recorded there.

Coleoptera

Dicladispa armigera (Coleoptera: Chrysomelidae)

Rice hispa

Distribution

Bangladesh, Burma, southern China, India, west Malaysia, Nepal, Pakistan, Sumatra, Thailand, West Irian (CIE Map 228).

Symptoms

Serious economic losses are caused by this pest particularly in Bangladesh. The main attack is on young rice, both by adults and larvae.

Adult beetles feed on the green portion of the leaf leaving only epidermal membranes. Feeding damage shows as characteristic white streaks along the long axis of the leaf. Larvae mine into the leaf between the epidermal membranes producing irregular longitudinal white mines. The damage starts from oviposition sites near the leaf tip and extends towards the base of the leaf blade. Affected leaves wither and die.

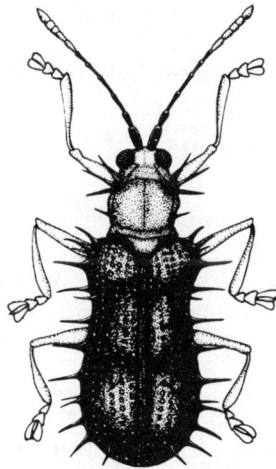

Fig. 69. Rice hispa, *Dicladispa armigera*. (H.R.B.)

0   2mm

Life cycle

The adult (Fig. 69) is a small shiny, black beetle 5.5 mm long with spines over the body surface. Females lay an average of 55 eggs singly, partly inserted beneath the epidermis of the ventral surface of rice leaves. The minute eggs are usually found towards the leaf tip, partially covered by a dark secretion from the female, they hatch in 4—5 days.

Pale yellow larvae, dorso-ventrally flattened and 2.4 mm long hatch and immediately burrow into leaf tissue, where they feed for 7—12 days and grow to 5.5 mm before pupation. The pupae are also flattened, brown and exarate. The pupal stage lasts 4—5 days, when adult beetles emerge from the larval tunnel and begin to lay eggs 3—4 days later.

Dicladispa gestroi (Coleoptera: Chrysomelidae)

Distribution

Malagasy Republic (CIE Map 206).

Symptoms

Adults eat the leaf surface while larvae mine the leaves, their attacks result in withered leaves with white patches. This species is a pest of rice in the coastal regions and is as important a pest as the stem borers in this area.

Dicladispa viridicyanea (Coleoptera: Chrysomelidae)

Distribution

Burundi, Kenya, Zaire.

Symptoms

This insect is suspected of transmitting rice yellow mottle virus in Kenya (p. 84) and is recorded as damaging rice in Zaire and Burundi. The larvae mine the leaves making transparent patches, adults eat leaf tips, the effect being particularly important when the rains are late or less heavy than usual.

Life cycle

The adult beetle is a metallic blue-green, about 5 mm long with 5 lateral spines on each side of the thorax and a series of lateral spines on the elytra, alternately long or short. The female lays eggs on the under surface of the young leaves and covers them with excrement. There are 3 to 6 generations annually, the life cycle takes 24—40 days to complete.

Trichispa sericea (Coleoptera: Chrysomelidae)

African rice hispa

Distribution

Angola, Burundi, Cameroun, Ethiopia, Ivory Coast, Kenya, Malagasy Republic, Mali, Rwanda, Senegal, Sudan, Swaziland, Tanzania, Togo, Uganda, Zaire (CIE Map 257).

Symptoms

Both adults and larvae (Fig. 70) feed on leaf tissues of young rice, the leaves become wafer thin and bleached. The first attack is very localised with an eventual rapid spread. A severe attack kills the plant. This species is recorded as able to transmit rice yellow mottle virus in Kenya (p.84).

Life cycle

White boat-shaped eggs 1 mm long are laid singly on the lower leaf surface. The larvae mine and pupate within those portions of the leaf lamina which are not submerged.

As leaves harden they become unattractive as adult food and for oviposition. The pest then migrates to other plants e.g. *Echinochloa holubii* a grass typical of drainage canals around rice fields. It has also been recorded on *Eragrostis heteromera, Eragrostis aethiopica, Digitaria zayheri, Diplachne fusca* and *Chloris virgata*, in Swaziland, but rice is the preferred host plant.

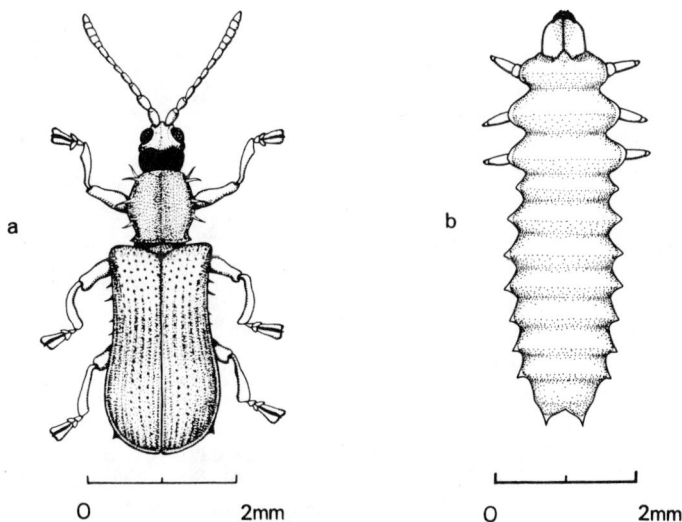

Fig. 70. Rice beetle, *Trichispa sericea,* (a) adult, (b) larva. (H.R.B.)

181

Control of hispas

Cultural

Keep bunds or dyke walls free from grassy weeds on which these beetles can maintain their populations. Avoid rattoon rice by ploughing in after harvesting. In some countries it is recommended that up to 15 cm of leaf tip should be clipped off and fed to cattle or destroyed in order to dispose of the eggs and larvae.

Chemical

See Table 15.

TABLE 15. RECOMMENDATIONS FOR CHEMICAL CONTROL OF HISPAS

Insecticide	Formulation	Active ingredient per hectare	Application
BHC	dust	0.84 kg	—
Diazinon	—	1.00 kg	—
Dichlorvos	—	0.3–0.5 kg	—
Dicrotophos	—	0.38 kg	—
Fenitrothion	—	0.6 kg	—
Fenthion	—	0.6 kg	—
Mephosfolan	—	0.5–1 kg	—
Phosmet	w.p., dust	—	Follow local recommendations
Phosphamidon	—	0.2–0.3 kg	—

Oulema oryzae (Coleoptera: Chrysomelidae)

Rice leaf beetle

Distribution

China, Japan, Korea, Manchuria, Ryuku Is., eastern Siberia, Taiwan.

Symptoms

Adults and larvae remove strips of the leaf surface causing a scorched appearance. Adults are metallic blue with head and antennae black and thorax yellowish brown. Larvae are dirty yellow with dark brown nodules and legs, each larva is covered in a crust of its own excreta. The percentage of yield reduction is 0.25 of the percentage of injured leaves to 50 hills (at random) in Japan.

Life cycle

Only one generation is produced during a year. The cylindrical eggs are laid in masses on the upper leaf surface, after 5—11 days larvae emerge, these develop for 13—19 days and then pupate in a whitish cocoon on the leaves (below ground in upland rice). Adults hibernate 3 months of the year in grass or weeds.

Control

Chemical

See Table 16.

TABLE 16. RECOMMENDATIONS FOR CHEMICAL
CONTROL OF *OULEMA ORYZAE*

Insecticide	Formulation	Active ingredient per hectare	Application
Azinphos-methyl	—	0.3—0.5 kg	—
Cartap	—	6 kg	—
DDT	dust	15 kg	—
Fenitrothion	u.l.v.	0.5 *l*	—
Fenthion	—	0.6—0.12 kg	—
Methomyl	—	0.5—0.75 kg	—
Phosmet	follow local recommendations		
Pyridaphenthion	dust	0.6 kg	—
Trichlorphon	—	0.8—1 kg	—

Chaetocnema concinnipennis (Coleoptera: Chrysomelidae)

Distribution

India.

Symptoms

Feeding by the beetles produces short straight lines on the leaf surface similar to the damage caused by the rice hispa *Dicladispa armigera*. Seriously affected parts of the field appear scorched due to withering of the damaged leaf tips. The beetles are members of the sub-family Halticinae which have swollen hind femora associated with a highly developed jumping ability. Slight disturbance of the crop causes the beetles to jump off the leaves.

Control

Cultural

Efficient control of weeds between rice crops will keep the population of these insects low.

Chaetocnema pulla (Coleoptera: Chrysomelidae)

Distribution

Kenya.

Symptoms

This beetle feeds in the same manner as *C. concinnipennis* eating narrow strips of the leaves parallel to the veins. This species is important because it has been shown to transmit rice yellow mottle virus (p.84) in Kenya, this disease is characterised by stunting, reduced tillering and crinkling, mottling and yellowish streaking of the leaves.

Control

Cultural

It is most important to maintain high standards of crop hygiene to cut down sources of the disease. Ideally harvested fields should be ploughed immediately if rice yellow mottle virus has been present. Grass growing on the bunds supports populations of *C. pulla* and where possible bunds should be kept clean.

Sesselia pusilla (Coleoptera: Chrysomelidae)

Distribution

Mozambique north to Kenya.

Symptoms

This beetle transmits rice yellow mottle virus (RYMV) the symptoms of which are described on p.84. The beetle causes little damage by feeding.

Control

Cultural

Keep the grassy weeds cleared as these are alternative hosts for the beetle and the virus.

Acarina (mites)

Oligonychus oryzae (Tetranychidae)

Paddy mite

Distribution

India.

Symptoms

Plants show characteristic whitish patches on the leaf surface, mites and their webs are present. When the attack is severe leaves turn greyish white and dry up.

Life cycle

The entire life cycle can take 8–12 days. The adults are sexually mature when they emerge from the deutonymph and mate as soon as possible. Egg laying begins in 1–3 days. Flattened spherical straw coloured eggs are laid singly in rows along leaf midribs and veins. Incubation lasts 4–9 days. This is followed by the three immature stages — larva, protonymph and deutonymph. This mite is active and breeds throughout the year building up a large population during hot weather when the life cycle takes the shortest time.

Control

Chemical

Carbophenothion, phorate or malathion at 0.03% are recommended.

STEM AND ROOT FEEDERS

Orthoptera

Gryllotalpa africana (Orthoptera: Gryllotalpidae)

Mole cricket

Distribution

Africa, tropical Asia, Europe, Japan (CIE Map 293).

Symptoms

Both nymphs and adults (Fig. 71) burrow along beneath the soil surface and attack the stems of rice plants (and other plants) below ground and close to the roots. The stem is usually cut through and the plants die. Damage is localised to the area of the insects' burrowing, but large numbers of rice plants can be destroyed during a night. Young and newly planted seedlings are most commonly attacked in the early part of the season before flooding has taken place.

Upland rice is more heavily attacked when the fields are damp.

Life cycle

Adult crickets are large, light brown insects, 25–35 mm in length. They are characterised by their greatly enlarged and modified forelegs for burrowing into soil around plant roots. Adults cannot live in the rice fields after flooding, but swim to the bunds, where the females burrow into the soil and construct hardened cells in which the eggs are laid. Each cell usually contains 30–50 eggs

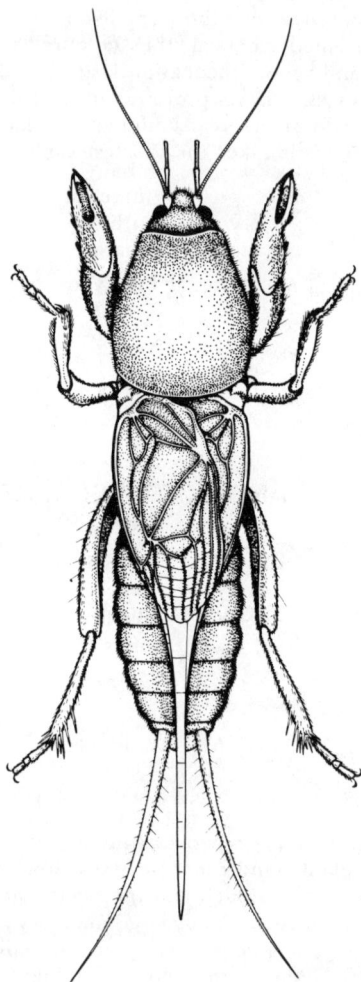

Fig. 71. Mole cricket, *Gryllotalpa africana.*
(H.R.B.)

and these take 15—40 days to hatch depending on the temperature. Nymphs hatch and feed on roots, causing bare patches in the field. The nymphal period lasts 3—4 months and there is usually only one generation per year. The overwintering period is spent by the adult in burrows deep in the soil.

Control

Chemical

Mole crickets can be destroyed with a poison bait. One part of sodium fluosilicate and one part of brown sugar are mixed with ten parts of carrier (rice bran is effective) and mixed to a stiff paste with water. Cakes of this after drying, can be distributed where damage is apparent. Populations can be kept down by adults being collected when fields are ploughed.

Hemiptera: Homoptera

Rhopalosiphum rufiabdominalis (Hemiptera — Aphidae)

Rice root aphid, red rice root aphid

Distribution

Wide distribution which includes most rice growing areas (CIE Map 289).

Symptoms

Plants wilt and die, large numbers of aphids are found on the upper parts of the roots. Usually the effect of the aphids feeding is not so extreme, plants only yellow and become slightly distorted.

Life cycle

In Taiwan the aphid has 53—58 generations per year. Both alate and apterous females reproduce. The apterous female is dark or light green with large red blotches and the alate form is mainly reddish brown. Completion of the nymphal instars takes about 7 days. One to two days after becoming adult the aphid begins to reproduce.

Spring migrants (alate forms) fly from overwintering host plants such as other Gramineae, tobacco or *Prunus* spp. to infest rice. During the rice growing season the majority of aphids produced are apterous and remain on rice. Just as the crop is maturing alatae increase in number and disperse to alternative hosts. In Japan the most important host of this aphid is the cherry tree.

This species is recorded associated with ants on the roots of various hosts.

Control

Cultural

Late sowing and the application of ammonium sulphate or farmyard manure are useful.

Chemical

This pest is important in upland rice in Japan but elsewhere apparently not of economic importance. The application of phosphamidon at 0.2–0.3 kg a.i./ha or diazinon at 0.4 *l* a.i./ha should control this pest.

Minor aphid pests

Sipha glyceriae is a green aphid found on rice in Italy where it is parasitised by *Aphelinus asychis*. If it increases in numbers to pest status fenthion, fenitrothion or malathion at the manufacturers recommended dosage will control it.

Schizaphis graminum (CIE Map 173), the greenbug which is found in China, Japan, India, Near East, USA, South America and USSR feeds on wheat, barley, oats and Italian millet in addition to rice. In USSR it causes damage to rice in Central Asia and the Ukraine. It overwinters in the egg stage on the shoots of winter and wild cereals, later moving to rice. In an outbreak in Krasnador 2.5% parathion-methyl dust at 10 kg/ha gave effective control.

Mealybugs (Hemiptera: Pseudococcidae)

Several species of mealybug have been recorded on rice, living on roots, stems and leaves. Records indicating that this group of insects are causing damage of economic importance are unusual and the presence of mealybugs without serious symptoms in the plants ought not to be taken to indicate a need for the application of a pesticide.

Dysmicoccus boninsis (Hemiptera: Pseudococcidae)

Distribution

A very widespread pest of sugarcane which has been recorded on rice in Formosa (CIE Map 116).

Symptoms

Elongate oval greyish mealybugs 4 mm long (oldest specimens) with about six short lateral wax filaments on the abdomen. They are found on the leaves. On sugarcane they are most abundant beneath the bases of the leaf sheaths.

Dysmicoccus brevipes (Hemiptera: Pseudococcidae)

Distribution

A tropicopolitan species probably found wherever pineapple is grown. Recorded on rice in Brazil, New Guinea, Paraguay and Zaire (CIE Map 50).

Symptoms

These mealybugs attack the roots of rice and may be accompanied by ants.

Geococcus oryzae (Hemiptera: Pseudococcidae)

Distribution

Japan and North Korea.

Symptoms

Small (2.0 mm) apterous insects covered by a white cottony secretion, feeding on the underground stems of grasses and roots of rice. No damage to rice has been reported.

Heterococcus rehi (Hemiptera: Pseudococcidae)
Rice mealybug

Distribution

Bangladesh, India, Java, Nepal, Pakistan.

Symptoms

Pink mealybugs, elongate oval to broadly oval (5 mm long) often living in masses between sheath and stem. An infested field has isolated patches of sickly stunted scorched looking plants. Damage is intense and when there is a heavy infestation ears may not be formed or may not emerge.

Life cycle

Reproduction is parthenogenetic, oviviparous and viviparous. The offspring are protected, sometimes in an ovisac of waxy threads and at others under the body of the female, or between wax plates secreted from the end of the abdomen. Nymphs crawl out of the egg sac and move, or are wind blown, to upper parts of the same or nearby plants. The nymphs become fixed between the leaf sheath and stem by driving their proboscises into the stem of the rice plant.

A single female can produce 60–200 eggs or nymphs. The larval stage lasts 15 days. Mature females do not move, but males fly off. The entire life cycle lasts 17–37 days and there may be 12 generations in a year. The pest can survive on graminaceous weeds between crops of rice. It prefers plants growing in swampy areas or standing water.

Planococcoides lingnani (Hemiptera: Pseudococcidae).

Distribution
China, Java, Malaysia.

Symptoms
Recorded on rice stems, the external appearance has not been described.

Control of mealybugs
Chemical

Diazinon at 0.3—0.6 *l* a.i./ha, dichorvos at 0.3—0.5 kg a.i./ha, dimethoate at 1 *l* a.i./ha, phorate at 1—2 kg a.i./ha or phosmet as recommended by the manufacturers should control mealybugs.

Diptera
Chironomus spp. (Diptera: Chironomidae)
Cricotopus spp. (Diptera: Chironomidae)

Symptoms
Germinating seedlings collapse and die due to larvae feeding on primary roots and shoots. Seedlings are sometimes uprooted. In Egypt a *Chironomus* sp. is a problem in saline soils. The Australian *Chironomus tepperi* prefers soils with a high organic content following sod seeding. *Chironomus thummi, C. cavazzai, Cricotopus sylvestris* and *C. trifasciatus* are pests of European rice seedlings. In USSR a *Chironomus* sp. is reported feeding on floating or submerged rice leaves causing a severe reduction in shoot numbers.

Control
Cultural

Pregerminating seeds before sowing is recorded as leading to reduced losses. Draining the field 3—5 days after sowing kills the aquatic larvae. Care must be taken not to dehydrate the seedlings. When established plants were attacked in USSR lowering the irrigation water to 10 cm gave control.

Chemical

In Europe trichlorphon is recommended.

Diopsis thoracica (Diptera: Diopsidae)
Stalk-eyed borer

Distribution

Tropical West and East Africa, Somalia, Zanzibar.

Symptoms

Dead hearts. This pest (Fig. 72) causes the greatest loss of rice in the valley of the Benue in North Cameroun and is also recorded as damaging rice in Swaziland and Sierra Leone. It has not been found to attack plants other than cultivated rice, but as the adults occur in vast number where the humidity is high it seems probable that it has alternative hosts among wild rices and grasses, though these have not yet been discovered.

Life cycle

Adults are typical diopsids easily recognised by their characteristic eyes borne on the end of stalks. They always live near water and prefer aquatic plants and a shady habitat. *D. thoracica* has a characteristic red colouration on the abdomen. Females lay eggs singly on the upper surface of young leaves, normally in the channel formed by the midrib. Eggs are rarely laid on lower leaves, the underside of leaves or stems. They are fixed to the leaf with a cement which prevents their being washed off in heavy rains. Eggs are approximately 1.7 mm by 0.4 mm. They are boat-shaped, striated and have a characteristic anterior projection. When laid they are creamy-white but they later darken to tan. Each female adult lays about 20 eggs over a 10 day period.

The emerged larva (Fig. 73) moves down the inside of the leaf sheath and feeds above the meristem on the central spindle of young leaves causing the dead heart symptom. Later generations of larvae feed on the developing flower head. Larvae are about 18 mm long and 3 mm wide, whitish-cream with yellow markings on the terminal segments and have very small heads. The larval stage lasts 25—33 days.

Pupae are reddish in colour with brown dorsal bands and a transparent white undersurface. They are fat and almost triangular in section because of their compression inside the stems. During this stage the imago can be seen developing inside the pupal cases. Adults emerge after a 10—12 day pupal period.

Control

Control methods directed towards the adult are not considered practical. The most successful technique would seem to be the use of some form of systemic insecticide against the younger larval stages. However, the economic importance of this insect has not yet been determined and insecticidal application may not be justified.

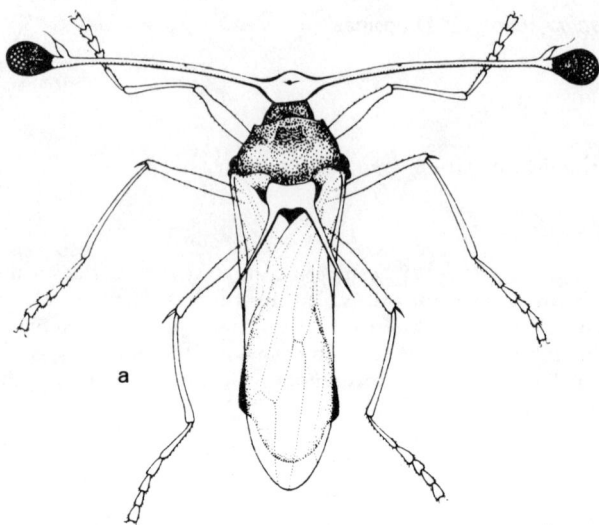

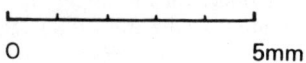

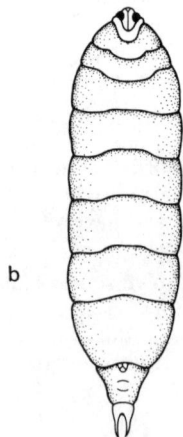

Fig. 72. Stalk-eyed borer, *Diopsis thoracica,* (a) adult, (b) pupa, ventral view. (H.R.B.)

Fig. 73. Dipterous stem borer larvae feeding inside a rice stem. (Shell Photographic Service)

Chlorops oryzae (Diptera: Chloropidae)
Rice stem maggot, rice shoot fly

Distribution

Indonesia, Japan.

Symptoms

This is a pest of graminaceous plants and attacks wheat, barley and rye as well as its preferred host, rice. Larvae bore into the stem near the growing point and feed on leaf blades, leaving broad chewed patches along the margins. On occasions they also cause punctures in the leaf. Tillering is reduced and stunting results from early attacks. Later generations feed on developing grains before the panicle emerges and this reduces the yield per panicle.

Life cycle

There are two distinct groups of this pest in Japan. In the north there are two generations and in the south three generations per year. The first broods of adults emerge in May. They look like small houseflies. Adults live for two weeks during which the females lay 50—100 small, white, elongate eggs singly on leaf blades of rice seedlings in the seedbed.

First generation larvae hatch in about one week. The larvae are about 1 mm long, translucent and white. They migrate to the central whorl of rice and begin feeding. The larval stage lasts about 6 weeks, then pupation, which lasts 2 weeks, occurs between the leaf sheath and stem. Later broods oviposit on grasses where the larvae overwinter.

Control

Cultural

Some varieties of rice are less susceptible to damage by this pest, in areas of heavy infestation it is wise to plant these varieties if they are suitable in other respects.

Chemical

The application of dimethoate or diazinon granules at 1.5 kg a.i./ha, fenthion dust at 0.6—1.2 kg a.i./ha or phorate at 1—2 kg a.i./ha have been recommended.

Orseolia oryzae (Diptera: Cecidomyiidae)

Rice gall midge; maleng bug

Distribution

Tropical Asia except the Philippines and Malaysia, Nigeria, northern Cameroons, Sudan (CIE Map 171).

Symptoms

Tillers become silvery and thickened (Fig. 74) due to the development of a hollow tube-shaped gall, this symptom is known as 'onion leaf' or 'elephant tusk' (Plate 3).

Life cycle

The fly (Fig. 75) requires a high humidity for activity and passes the dry season on grasses or wild rice in swampy areas. When the rainy season starts growth among wild hosts begins and the stem gall midge population multiplies so that when rice is planted it is attacked. Rice is susceptible during the vegetative phase of its growth, that is until tillering stops and formation of the panicle (booting stage) begins. An entire life cycle on rice takes 9—26 days and slightly less on some wild grasses.

Adult males are about 3 mm long with yellowish-brown bodies. The females are slightly larger, up to 3.5 mm long. Their bodies are bright reddish brown. Adults live from 1—5 days during which time they only feed on water on the plant surface. Fertilised females start egg-laying within a few hours of emergence. Their egg-laying capacity varies from 100—300 eggs. The eggs are elongate, tubular 0.5 mm long and 0.125 mm wide. Their colour varies from white or pink to red, or even yellow but all eggs become shining amber before hatching. Incubation lasts 3—4

Fig. 74. Gall caused by *Orseolia oryzae,*
the entire tiller has been transformed into
typical 'onion shoot'. (T. Hidaka)

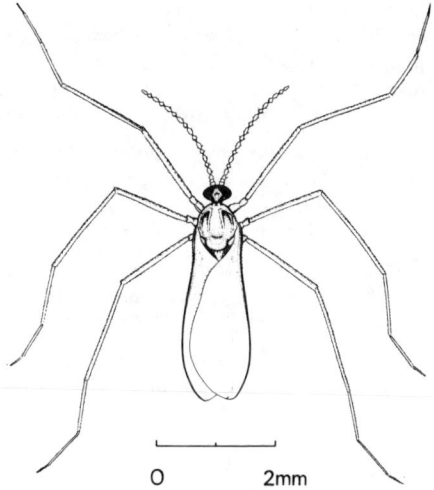

O   2mm

Fig. 75. Rice gall midge, *Orseolia oryzae.*
(H.R.B.)

days or more depending on the relative humidity, 82–88% being the most
favourable.

The larva is 1 mm long on hatching, fairly stout with a pointed anterior end. It
grows to about 3 mm and its colour darkens from cream to pale red. Larvae, soon
after hatching, move down between the leaf sheaths until they reach the growing
point of the apical or side buds, this takes 6–12 h on a 2 week old rice seedling.
If the humidity is not high enough to cause condensation on the leaf surface the
larvae cannot move, their bodies desiccate and they die if a high humidity is not
restored within 24 h. Once at the apical shoots they feed at the base of the
growing point. Due to their presence a gall forms and panicles do not develop in
infected tillers. Pupation takes place after 15 to 20 days.

The pupa remains at the base of the gall, it is 2.0–2.5 mm long and 0.6–0.8 mm
wide, pale pink at first, but becoming darker pink to red before the emergence of
the adult midge, 2–8 days after pupation. The pupa possesses several rows of
abdominal spines, which are subequal and point backwards, thus enabling the pupa
to wriggle up to the tip of the gall.

Before the adult emerges the pupa makes a hole in the top of the gall with its
spines and projects half way out; its skin then splits and the adult emerges, usually
during the night.

Control

Cultural

Generally it is recommended that rice should be planted as early as possible in relation to the rains so that when the pest moves from wild hosts into the rice, the plants will be well advanced in growth. Some work in Thailand has shown that rice planted later suffered less from this pest because as the season progressed parasites increased in number and the incidence of attack went down.

Resistant varieties of rice do exist but they are of poor plant type and grain quality. Attempts are being made to incorporate the resistance characteristic into useful varieties and these will be available in the future. The use of quick growing varieties is frequently recommended to prevent serious attacks.

One worker has found that a high potash application will encourage gall midge attack. Fertilizer applications should be no greater than the local recommendations suggest.

The removal of the alternative hosts, weeds such as: *Cynodon dactylon*, *Pseudhomonylia fluitans* and *Panicum fluitans* or wild rice species e.g. *Oryza barthii*, during the dry season may assist in cutting down the pest population.

Chemical

Timing of the application of insecticides is important, killing the adults or first instar larvae before they have reached the site of the gall is the aim. The vulnerable stage of the rice plant is 20–45 days after transplanting and farmers, by observing galls in weeds and rice, should note when the adults have emerged and spray at that time. See Table 17 for the chemicals recommended.

TABLE 17. RECOMMENDATIONS FOR CHEMICAL
CONTROL OF *ORSEOLIA ORYZAE*

Insecticide	Formulation	Active ingredient per hectare	Application
Azinphos-methyl	–	0.3–0.5 kg	–
Carbofuran	–	0.45–0.6 kg	–
Diazinon	–	1.5 kg	Diazinon impregnated tapes are also recommended
Fenitrothion + cyanofenphos	e.c.	1.15 l (formulation)	Dilute 800–1000 times with water
Mephosfolan	gran.	0.5–1 kg	–
Phorate	gran.	1–2 kg	–
Phosphamidon	–	0.4–0.5 kg	–
Phosmet	follow local recommendations		–
Salithion	gran.	0.5–1 kg	–

Atherigona spp. (Diptera: Muscidae)
Rice seedling fly, bibit fly, stem-mining maggot

Distribution

India, Indonesia, Japan, Malaysia, New Guinea, Philippines, Sri Lanka.

Symptoms

These insects are also pests of wheat, maize and many grass weeds. On rice they produce typical symptoms of twisted whitish dead hearts similar to those caused by lepidopterous stem borers. The seedlings of upland rice are most commonly attacked. Larvae feed on the central shoot which turns yellow and dies.

Life cycle

Adults resemble small houseflies (Fig. 76). They are about 3 mm long with a grey thorax, yellow spotted abdomen and yellow legs; they have a very distinctive angular head with deep set antennae.

Eggs are laid on the upper and lower surfaces of rice seedling leaves, most commonly in wet weather late in the day. When the tiny larvae hatch they move down the leaf in the water film, get between the leaf sheath and stem into the growing point, and they feed on the basal part of the youngest leaf. Pupation takes place in the soil. The complete life cycle takes 15—32 days.

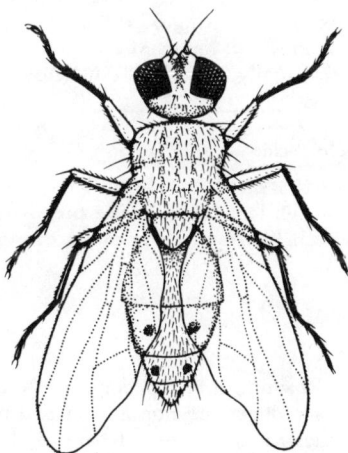

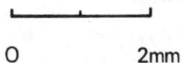

Fig. 76. Rice seedling fly or stem-mining maggot, *Atherigona oryzae*. (H.R.B.)

O   2mm

Control

Cultural

Flooding for 24 h to kill the pupae may be helpful. This treatment should be repeated after several days.

Chemical

Phosphamidon at 0.2–0.4 kg a.i./ha is recommended, also dusting nurseries with 5% BHC daily during the first week after germination if infestation is very high. Granular systemic insecticides are effective in nurseries. Insecticidal seed dressings may reduce the incidence of this pest.

Coleoptera

Heteronychus oryzae (Coleoptera: Scarabaeidae)

Distribution

Nigeria, Sierra Leone.

Symptoms

Newly sown rice up to 6 weeks old is most commonly attacked, the first sign of damage being wilting of the central leaves, followed by progressive wilting of the outer ones and death of the plant. Both adults and larvae attack rice stems and roots, usually about 2.5 cm below ground level.

Life cycle

The adult is a small (9–10 mm) dark reddish-brown to black beetle with reddish-brown legs. The beetle breeds in piles of rotting weeds and grass. The larvae which feed on young rice are small, grub-like, and white in colour.

Control

Cultural

Better field sanitation, especially the removal of breeding sites in piles of grass and weeds reduces the population of this pest. If bunds are kept weed free and stubbles destroyed the population of the beetles may be so reduced that chemical control is not necessary.

Another *Heteronychus* sp., *H. arator* (syn *H. sanctaehelenae* CIE Map 163) has been noted causing similar damage to rice in South Africa and east Australia.

Colaspis flavida (Coleoptera: Chrysomelidae)
Grape colaspis

Distribution
Southern USA (Mississippi basin).

Symptoms
Damage is caused by overwintering larvae feeding on germinating rice seeds or seedlings. The stand of rice may be drastically reduced. Reduction in tillering is also a result of larval attack. This pest is most common in dry-seeded fields.

Life cycle
This pest also attacks the forage legume lespedeza and other leguminous crops which are its preferred hosts. Adult beetles (Fig. 77) lay their eggs in the soil around the roots of leguminous plants. Larvae hatch and spend up to 7 months in cells pressed into the earth above or slightly into the subsoil. As rice is planted in the rotation larvae feed on their roots, seeds and seedlings. Pupation then occurs and the pupae remain in the soil for 3—7 days until the adults emerge.

Control
Chemical

Rice planted following a lespedeza crop should be seed treated with insecticide. Chlorpyrifos at 0.56—1.12 kg a.i./ha has shown promise in field trials in controlling this pest but it is not currently registered for use on rice.

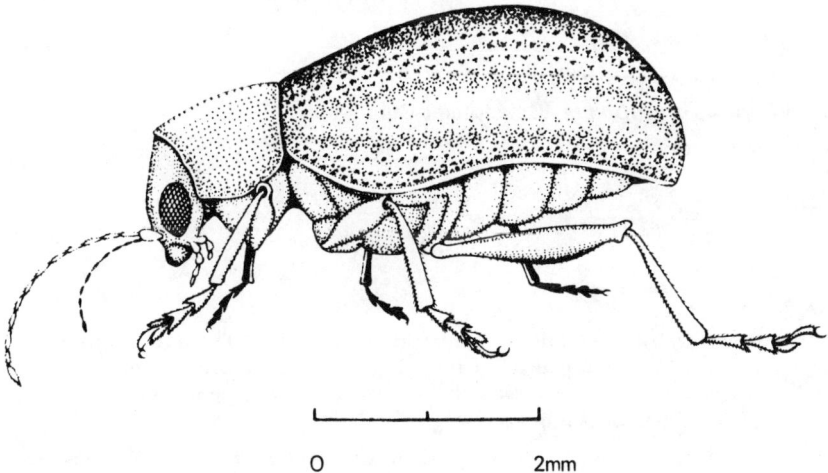

O   2mm

Fig. 77. Grape colaspis, *Colaspis flavida*. (H.R.B.)

Echinocnemus oryzae (Coleoptera: Curculionidae)
Rice root weevil

Distribution
Southern India

Symptoms
The plants do not thrive, they are stunted and gradually turn yellow, tillering is very poor and some plants may die. Investigation of the roots reveals the beetle larvae adhering to them. Clay or heavy loam soils favour this pest.

Life cycle
The dark adult weevils, which are covered in grey scales, emerge in June and can be seen swimming in the water or moving about the rice. Eggs are laid on rice and other Gramineae. On hatching the larvae feed on the stem epidermis and then go into the soil to attack the roots and when mature pupate about 8 cm below soil level. Late in the season larvae go 30 cm below soil level where they remain until the next year. There are two generations per year.

Control
Cultural
Encourage vigorous growth of the rice.

Chemical
Dip the seedlings for 15 min in 0.1% diazinon solution.

Lissorhoptrus oryzophilus (Coleoptera: Curculionidae)
Rice water weevil

Distribution
Rice growing regions of the USA (CIE Map 270).

Symptoms
This pest attacks rice in both its adult and larval stages. The adult (Fig. 78a) feeds on leaves of young plants leaving a long scar where the leaf surface has been removed. Damage caused by the adult is not of any importance when compared with that of the larvae.

The larvae (Fig. 78b) feed within and upon rice roots. When populations are high root systems may be so reduced that entire plants lodge and are uprooted. Infestation causes stunting, delayed maturity and loss of grain yield.

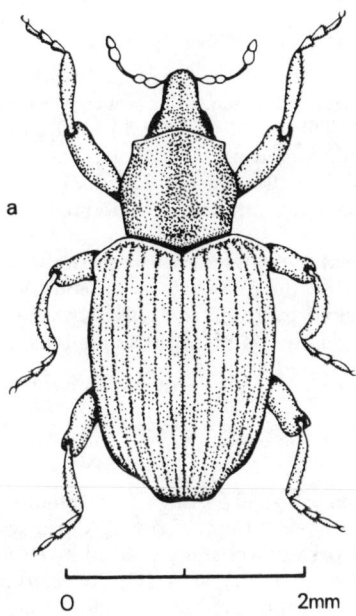

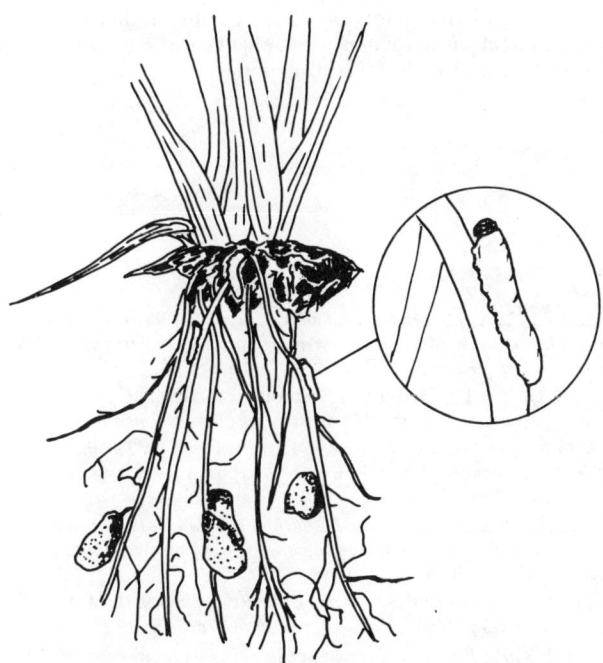

Fig. 78. Rice water weevil (a) adult, (b) larvae and cocoons (COPR). (H.R.B.)

Life cycle

Semi-aquatic adults fly into the standing crop and move from plant to plant by swimming just below the water surface. The adult is 3.0 mm long, greyish with a darker area on its back. The female moves down the rice stem and lays eggs on the basal half of the submerged portion of the leaf sheath and, rarely in the roots. Oviposition begins after the rice is flooded. Eggs are white and cylindrical and hatch under field conditions in about 8 days. Newly hatched larvae feed for a time in the leaf sheath and then go down to the roots, emerging from the leaf sheath and falling through the water to the soil and roots. The pupa is white and the same size as an adult weevil. It is enclosed in a cocoon which is attached to a rice root. Newly emerged adults usually fly to adjacent fields of younger rice at night. The adults spend the overwintering period in rice stubbles or matted grass.

Control

Chemical

The application of insecticides usually brings an economic return when the larval population exceeds 25 per 30 cm of row. Aldrin seed dressings formerly controlled this pest but it became resistant to aldrin. Other insecticides have been tested and chlorpyrifos (0.3–0.5 kg a.i./25 kg seed) and pirimiphos-ethyl (0.1–0.2 kg a.i./ha) were the only effective larvicides that did not significantly reduce the number of rice plants or interact with the herbicide propanil to produce seedling leaf burn. When propanil has not been used carbofuran granules (0.5 kg a.i./ha) or bufencarb granules (1 kg a.i./ha) may be applied 2–7 days after flooding.

Bibliography

ABUL-NASR, S., ISA, A. L. and EL-TANTAWY, A. M. (1970). Control measures of the bloodworms, *Chironomus* sp. in rice nurseries (Diptera). *Bulletin of the Entomological Society of Egypt, Economic series* 4: 127–133.

AHMAD, I. (1965). The Leptocorisinae (Heteroptera: Alydidae) of the world. *Bulletin of the British Museum (Natural History) Entomology Supplement* No. 5.

ALAM, M. Z. (1964). Insect pests of rice in East Pakistan. In *The major insect pests of the rice plant.* International Rice Research Institute. pp. 643–655. Johns Hopkins Press, Baltimore, 1967.

APPERT, J. (1970). Etudes et travaux, *Maliarpha separatella* (borer blanc du riz) Observations nouvelles et rappel des problèmes entomologiques du riz a Madagascar. *Agronomie Tropicale* 25(4): 329–367.

APPERT, J., BETBEDER-MATIBET, M., and RANAIVOSOA, H. (1969). Twenty years of biological control in Madagascar. *Agronomie Tropicale* 24 (6–7): 573–585.

ASAKAWA, M. (1975). Insecticide resistance in agricultural insect pests of Japan. *Japan Pesticide Information* 23: 5–8.

BAKKER, W. (1970). Rice yellow mottle, a mechanically transmissible virus disease of rice in Kenya. *Netherlands Journal of Plant Pathology* 76: 53–63.

BAKKER, W. (1971). Three new beetle vectors of rice yellow mottle virus in Kenya. *Netherlands Journal of Plant Pathology* 77: 201–206.

BELLI, G., CORBETTA, G. and OSLER, R. (1975). Recherche e osservazioni sull'epidemiologia e sulle possibilita di prevenzione del giallume del riso. *Il Riso* 24(4): 359–363.

BOWLING, C. C., (1962). Effect of insecticides on rice stink bug populations. *Journal of Economic Entomology* 55(5): 648–651.

BOWLING, C. C. (1963). Cage tests to evaluate stink bug damage to rice. *Journal of Economic Entomology* 56(2): 197–200.

BOWLING, C. C. (1964). Insect pests of rice in the United States. In *The major insect pests of the rice plant*. International Rice Research Institute. pp. 551–570. Johns Hopkins Press, Baltimore, 1967.

BOWLING, C. C. (1968). Rice water weevil resistance to aldrin in Texas. *Journal of Economic Entomology* 61(4): 1027–1030.

BOWLING, C. C. (1969). Estimation of rice stink bug populations on rice. *Journal of Economic Entomology* 62(3): 574–575.

BOWLING, C. C. (1972). Note on the biology of rice water weevil *Lissorhoptrus oryzophilus*. *Annals of the Entomological Society of America* 65(4): 991.

BRENIÈRE, J., RODRIGUES, H. and RANAIVOSOA, H. (1962). Un ennemi du riz à Madagascar. *Maliarpha separatella* Rag. ou borer blanc. *Agronomie Tropicale* 16: 223–302.

BRITTON, E. B. (1959). *Heteronychus oryzae* sp. n. (Coleoptera: Scarabaeidae), a pest of rice in Sierra Leone. *Annals and Magazine of Natural History* (series 13). 2(15): 169–170.

DESCAMPS, M. (1957). Contribution a l'etude des diptères Diopsidae nuisables au riz dans la Nord-Cameroun. *Journal d'Agriculture Tropicale et de Botanique Appliquée* 4(1–2): 83–93.

DYCK, V. A. (1974). *Pest damage to plants and economic thresholds*. Paper presented at International Rice Research Conference 1974, IRRI, Philippines.

DYCK, V. A. (1974). *Recent advances in the integrated control of insect pests of rice*. Paper presented at International Rice Research Conference, 1974, IRRI, Philippines.

FAO (1971). *Crop loss assessment methods*. pp. 112. FAO, Rome and Commonwealth Agricultural Bureaux, Farnham Royal, UK.

GIFFORD, J. R., OLIVER, B. F., and TRAHAN, G. B. (1972). Insecticidal seed dressings on drill-seeded rice to control the rice water weevil. *Journal of Economic Entomology* 65(5): 1380–1383.

GRIGARICK, A. A. (1959). Bionomics of the rice leaf miner *Hydrellia griseola* (Fallen), in California (Dipter: Ephyd.). *Hilgardia* 29(1): 1–80.

GRIST, D. H. and LEVER, R. J. A. W. (1969). *Pests of rice*. pp. 520. Longmans, Green & Co., London.

HAMA, H. (1975). Resistance to insecticides in the green rice leafhopper. *Japan Pesticide Information* 23: 9–12.

HATTORI, I. (1971). Stem borers of graminaceous crops in southeast Asia. In *Symposium on rice insects, July 1971*. Tropical Agricultural Research Center, Tokyo, Japan. pp. 145–153.

HSIEH, C.–Y. (1970). The aphids attacking rice plants in Taiwan (II) Studies on the biology of the red rice root aphid, *Rhopalosiphum rufiabdominalis* (Sasaki) (Aphidae, Homoptera). *Plant Protection Bulletin, Taiwan* 12(2): 68–78.

IKEYAMA, M. and MAEKAWA, S. (1973). Development of Spanone for the control of rice stem borers. *Japan Pesticide Information* 14: 19—22.

INGRAM, J. W. (1967). Insects injurious to the rice crop in Washington. *Farmers Bulletin, US Department of Agriculture* 1543. pp. 17.

ISRAEL, P. and RAO, Y. S. (1961). Incidence of Gundhy bug and steps for its control. *Proceedings of the Rice Research Congress, Cuttack.* pp. 297—299.

ISELY, D. and SCHWARDT, H. H. (1934). The rice water weevil. *Bulletin of the Arkansas Agriculture Experiment Station* 299. pp. 44.

IWATA, T. (1970). High yield rice cultivation and the use of pesticides, B Insects. *Japan Pesticide Information* 5: 12—17.

IWATA, T. (1972). 1971 Evaluation of candidate pesticides (A—I) Insecticides: Rice. *Japan Pesticide Information* 12: 5—10.

IWATA, T. (1973). Rice insect control by fine granular formulation of insecticides in Japan. *Japan Pesticide Information* 14: 23—26.

IWATA, T. (1973). 1972 Evaluation of candidate pesticides (A—I) Insecticides: Rice. *Japan Pesticide Information* 16: 5—9.

IWATA, T. (1974). 1973 Evaluation of candidate pesticides (A—I) Insecticides: Rice *Japan Pesticide Information* 20: 5—8.

IWATA, T. (1975). 1974 Evaluation of candidate pesticides (A—I) Insecticides: Rice *Japan Pesticide Information* 24: 5—8.

KAPUR, A. P. (1964). Taxonomy of stem borers. In *The major insect pests of the rice plant.* International Rice Research Institute. pp. 3—43. Johns Hopkins Press, Baltimore, 1967.

KAZANO, H. (1974). Substitute insecticides after the restrictions on the organochlorine insecticides. *Japan Pesticide Information* 18: 22—26.

KOK, L. T. and PATHAK, M. D. (1967). Bioassay determination of gamma-BHC absorbed from soil and translocated in potted rice plants. *International Rice Commission Newsletter* 16(2): 27—34.

KOK, L. T. and VARGHESE, G. (1966). The four major lepidopterous rice stem borers in Malaya. *Malayan Agricultural Journal.* 45(3): 275—88

KUROSAWA, E. (1940). On the rice yellow dwarf occurring in Taiwan. *Journal of Plant Protection* 27: 61—66.

LING, K. C. (1972). *Rice virus diseases.* pp. 134. International Rice Research Institute, Los Banos, Philippines.

LING, K. C. (1973). *Synonymies of insect vectors of rice viruses.* pp. 29. International Rice Research Institute, Los Banos, Philippines.

MISRA, B. C. and ISRAEL, P. (1968). Studies on the bionomics of paddy mite *Oligonychus oryzae* (Hirst). *Oryza* 5(1): 32—37.

MITRA, D. K., RAYCHAUDHURI, S. P., EVRETT, T. R., GHOSH, A. and NIAZI, F. R. (1970). Control of the rice green leafhopper with insecticidal seed treatment and pre-transplant seedling soak. *Journal of Economic Entomology* 63(6): 1958—1961.

MOIZ, S. A., RIZVI, N. A. (1971). Ecological studies on *Tryporyza incertulas* (Walker) in southern part of West Pakistan. In *Symposium on Rice Insects, July 1971* Tropical Agricultural Research Center, Tokyo, Japan, pp. 19—26.

NASU, S. (1964). Rice leafhoppers. In *The major insect pests of the rice plant.* International Rice Research Institute. pp. 493—523. Johns Hopkins Press, Baltimore, 1967.

NYE, I. W. B. (1960). The insect pests of graminaceous crops in East Africa. *Colonial Research Studies* 31.

ODGLEN, G., and WARREN, L. O. (1962). The rice stink bug *Oebalus pugnax* Fabricius in Arkansas. *Arkansas Agriculture Experiment Station Report Series* No. 107. pp. 23.

OLIVER, B. F., GIFFORD, J. R. and TRAHAN, G. B. (1972). Evaluation of insecticidal sprays for controlling the rice stink bug in southwest Louisiana. *Journal of Economic Entomology* 65(1): 268–270).

OTANES, F. Q. and SISON, P. L. (1941). Pests of rice. *Philippine Journal of Agriculture* 12(2): 211–259.

PATHAK, M. D. (1964). Recent developments in and future prospects for the chemical control of the rice stem borer at IRRI. In *The major insect pests of the rice plant.* International Rice Research Institute. pp. 335–349. Johns Hopkins Press, Baltimore, 1967.

PATHAK, M. D. (1967). Significant developments in rice stem borer and leafhopper control. *PANS* (A) 13(1): 45–60.

PATHAK, M. D. (1969). Stem borer and leafhopper-planthopper resistance in rice varieties. *Entomologia experimentalis et applicata* 12: 789–800.

PATHAK, M. D. (1971). Resistance to leafhoppers and planthoppers in rice varieties. In *Symposium on rice insects, July 1971.* Tropical Agricultural Research Center, Tokyo, Japan. pp. 179–193.

PATHAK, M. D. (1972). Resistance to insect pests in rice varieties. In, *Rice Breeding.* International Rice Research Institute, Los Banos, Philippines.

PATHAK, M. D., ANDRES, F., GALAGAC, N. and RAROS, R. (1971). Resistance of rice varieties to striped riceborers. *International Rice Research Institute Technical Bulletin* No. 11.

PATHAK, M. D., BEACHELL, H. M. and ANDRES, F. (1973). IR20 a pest and disease resistant high yielding rice variety. *International Rice Commission Newsletter* 22(3): 1–8.

PATHAK, M. D. and DYCK, V. A. (1973). Developing an integrated method of rice insect pest control. *PANS* 19(4): 534–544.

PERERA, N. and FERNANDO, H. E. (1968). Infestation of young rice plants by the rice gall midge, *Pachydiplosis oryzae* (Wood-Mason) (Dipt. Cecidomyiidae), with special reference to shoot morphogenesis. *Bulletin of Entomological Research* 59: 605–613 (published 1970).

PEREZ, J. A. B. (1971). *La chinche del arroz.* pp. 24. Federacion Sindical de Agricultores Arroceras de Espana, Valencia.

PLANES, S., RIVERO, J. M. and MARTI FABREGAT, F. (1970). Estudios realizados sobre la plaga del barrenador del arroz (*Chilo suppressalis* Wlk = *simplex* Btlr). *Anales de Investigaciones Agronomicas* 19(3): 335–342.

PRAKASA RAO, P. S., KALODE, M. B., PRASANNA, K. DAS, VERMA, A. and ISRAEL, P. (1970). Field evaluation of new insecticides applied in standing water in the control of rice stem borers. *Oryza* 7(2): 113–120.

PRAKASA RAO, P. S., RAO, Y. S. and ISRAEL, P. (1970). Problems and prospects in the chemical control of rice stem borers. *Oryza* 7(2): 89–102.

RAI, B. K. (1973). *Spodoptera frugiperda* chemical control in paddy using ultra-low volume drift spraying. *Journal of Economic Entomology* 66(6): 1287–1289.

RAO, Y. S. and ISRAEL, P. (1964). Recent developments in, and future prospects for the chemical control of the rice stem borer in India. In *The major insect pests of the rice plant.* International Rice Research Institute. pp. 317–334. Johns Hopkins Press, Baltimore, 1967.

RIVERO, J. M. (1970). Empleo de insecticidas en los arrozales. *Anales del Instituto Nacional de Investigaciones Agronomicas* 19(3): 329–333.

RIVERO, J. M. (1971). Ensayo sobre las plagas animales del arroz en el area mediterranea y su control. *3 Journees de Phytiatre et de Phytopharmacie Circum-Mediterraneennes* Sassari (Italia).

RIVERO, J. M. and MARTI FABREGAT, F. (1965). Dos anos de experiencas de lucha contra el barrenador del arroz. (*Chilo suppressalis* Wlk). Investigaciones sobre la epoca conveniente para realizas los tratamientos. *Boletin de Patologia Vegetal y Entomologia Agricola* 28: 67–84.

ROLSTON, L. H. and ROUSE, P. (1965). The biology and ecology of the grape colaspis *Colaspis flavida* Say in the Arkansas Grande Prairie. *Arkansas Agriculture Experiment Station Bulletin* 694. pp. 31.

ROTHSCHILD, G. H. L. (1970). Some notes on the effects of rice ear-bugs on grain yields. *Tropical Agriculture* 47(2): 145–149.

SAKAI, M. (1971). The chemistry and action of cartap. *Japan Pesticide Information* 6: 15–19.

SOENARDI, I. R. (1964). Insect pests of rice in Indonesia. In *The major insect pests of the rice plant*. International Rice Research Institute. pp. 675–683. Johns Hopkins Press, Baltimore, 1967.

SRIVASTAVA, A. S. and SAXENA, H. P. (1964). Rice bug *Leptocorisa varicornis* Fabricius and allied species. In *The major insect pests of the rice plant*. International Rice Research Institute. pp. 525–548. Johns Hopkins Press, Baltimore, 1967.

SRIVASTAVA, A. S., GUPTA, B. P., AWASTHI, G. P. and SINGH, Y. P. (1967). Bionomics and survey of paddy root weevil. *Proceedings of the National Academy of Sciences India* 37(3)B: 233–234.

SUENAGA, H. (1974). Spray calendar for the control of diseases and insect pests of paddy rice crop in the southwest warm region of Japan. *Japan Pesticide Information* 21: 9–13.

TAMS, W. H. T. and BOWDEN, J. (1952). A revision of the African species of *Sesamia* Guenée and related genera, (Agrotidae – Lepidoptera). *Bulletin of Entomological Research* 43(4): 645–678.

BIRDS

Introduction

Rice probably suffers more from bird damage than any other tropical crop, with the possible exception of millet. Being small-grained both crops are attractive substitutes for the natural food of small seed-eaters, and may be taken at all stages of ripeness. Such seed-eaters live in enormous flocks in the dry tropics, where rice is becoming increasingly widely cultivated, and here the problem is far more acute than in the traditionally rice growing humid regions. It is not only the grain which is taken; rice also suffers considerable damage during the early stages of growth when the newly germinated seedlings are both eaten and badly trampled by migratory waterfowl, which are attracted in large numbers to inundated rice fields in otherwise dry regions.

Regional outline of pest problems

Africa

Rice is grown either under rainfed conditions where birds do little serious damage, or under flooded conditions. When the rice is inundated the areas under flood provide an extension of habitat for many species of resident waterfowl, and an ideal overwintering area for some migrant palaearctic species. The most important of the latter are the ruff *Philomachus pugnax*, black-tailed godwit *Limosa limosa*, and garganey *Anas querquedula*. All three feed extensively on newly planted seed, although the ruff and the godwit are known in Europe only as invertebrate feeders. Of the resident waterfowl those which are most abundant in rice fields are the spur-winged goose *Plectropterus gambensis*, Egyptian goose *Alopochen aegyptiacus* (Fig. 79) knob-billed goose *Sarkidiornis melanotos*, white-faced tree duck *Dendrocygna viduata* (Fig. 80) and fulvous duck *D. fulva*, all of which eat and trample large areas, as well as fouling the water so that it becomes unfit for rice growing.

The rice grain itself is taken by many species of seed-eaters which as a group do more damage than waterfowl. The principal pests in dry areas (Senegal, Tanzania, Swaziland) are the quelea *Q. quelea* (Figs. 81 and 86) and the golden sparrow *Passer luteus*, and in wetter areas (Sierra Leone, Congo, Southern Nigeria) the village weaver *Ploceus cucullatus* (Figs. 82 and 83). There are probably 50 or more other species which, whilst not causing widespread losses, may cause severe local damage to peasant-grown rice. Where irrigation is practised losses to birds are economically more important since the crop is being grown for cash and losses are at least more noticeable and more widely publicised.

Malagasy Republic

Waterfowl cause some damage but probably the most serious pest is a gregarious savannah weaverbird, the Madgascar fody *Foudia madagascariensis*.

Fig. 79. Egyptian goose *Alopochen aegyptiacus* a resident in African rice fields where it tramples large areas of growing rice. (A. S. Cheke)

Fig. 80. White-faced tree duck *Dendrocygna viduata*, this species eats young rice in Africa and also fouls the water in flooded fields making it unfit for rice growing. (A. S. Cheke)

Tropical Asia

In India, Burma and Sri Lanka some damage is done by weaverbirds of the genus *Ploceus*, principally the baya weaver *P. philippinus*, but the main losses are caused by parakeets, of which the most widespread and abundant species is the long-tailed parakeet *Psittacula krameri*. Numerous species of estrildids, (waxbills or grass-finches) may become pests locally throughout this region, but the most widespread is the spotted munia *Lonchura punctulata*. In Borneo three other species, *L. fuscans*, *L. malacca*, and *L. leucogastra* cause general losses but upland rice may suffer very severe local damage from the long-tailed munia *Erythrura prasina*.

In the Philippines four species are destructive to rice. *Lonchura ferruginosa*, *Lonchura punctulata cabanisi*, *Padda oryzivora* and the tree sparrow *Passer montanus*. The tree sparrow is a town dweller but forms large flocks and migrates to feed on rice during the harvest.

In Thailand *Ploceus* spp. cause severe losses as do sparrows in some upland areas where irrigated rice is grown at a time of year when no other grain source is available. Great trouble is taken to scare off the pests from the ripening grain, in some cases girls are employed from dawn to dusk to beat gongs and sling stones at the birds. It is doubtful if any economic reduction in grain losses results from these actions but in very bad areas of bird infestation it may be necessary to protect experimental rice breeding plots by the use of wire cages; this method is being used at IRRI in the Philippines.

Fig. 81. Quelea, (*Quelea quelea*) (a) male, (b) female. The most important bird pest of all food grains in drier areas of Africa. (A. S. Cheke)

Fig. 82. Village weaver, *Ploceus cucullatus* a typical seed-eating pest in wetter areas of Africa. (A. S. Cheke)

Fig. 83. Village weaver colony. Note the nests very close to water. (A. S. Cheke)

America

The most serious pests are some members of the family Icteridae, principally blackbirds *Agelaius* spp., but also the brown-headed cowbird *Molothrus ater* and the common grackle *Quiscalus quiscula*. $500,000–$1,500,000 is said to be lost each year in Arkansas alone as a result of blackbird damage. Probably the worst single offender is the red-winged blackbird *Agelaius phoeniceus* (Fig. 84). Before cultivation this species lived in large colonies in river valley swamps, many of which have now been reclaimed for rice growing. So that, although the natural feeding areas have been reduced the birds have been provided with an acceptable substitute. The flooding of the stubble fields for duck hunting has provided the blackbird with extensive alternative habitat for much of the year, and in 1959 even resulted in a second autumnal breeding season.

In the rice growing regions of Surinam an additional pest is the purple gallinule *Porphyrula martinica* (Fig. 85), which eats growing stems and tramples large areas.

Australia

Practically all the rice produced in Australia is in New South Wales in the Murrumbidgee and Murray river areas. There are no serious bird pests of rice in this area. If rice growing expands in the northern areas magpie geese, *Anseranas semperpalmata* which occur in very large numbers, could be very damaging. Serious damage was reported on a development scheme at Humpty-Do from magpie geese both feeding and trampling the young rice, control by shooting was attempted. Parrot species e.g. *Kakatoe sanguinea*, the little corella are a potential pest and were recorded feeding on rice both after sowing and during ripening and harvest. The spotted munia *Lonchura punctulata* has been introduced into Australia, it is spreading and may become a pest. The Java sparrow, *Padda oryzivora* has been introduced twice and liberated but failed to establish itself in the wild; it is, however a popular cage bird. In Western Australia the pest potential of this species is realised, and there is legislation to prohibit keeping the bird in captivity or importing it into Western Australia.

Possible control methods

One of three basic strategies may be adopted to avoid or reduce crop damage by birds;

1 The wholesale destruction of the pest species to bring its population down to a level where damage no longer occurs or is of no economic importance,

2 The use of bird scaring devices, chemical repellents, or 'bird-resistant' varieties to deflect the depredations of birds away from the crops, and,

3 The avoidance of bird damage by crop management.

Almost all measures used to date against bird pests in the tropics have fallen in category 1, and undoubtedly the greatest efforts have been directed towards the permanent reduction of the quelea population in Africa. Yet quelea is a classic example of a bird pest whose population can never be controlled by killing. The annual adult mortality is around 50%, probably due mainly to starvation in the

Fig. 84. Red-winged blackbird, *Agelaius phoeniceus* a pest in tropical America where it lives in swamps reclaimed for rice growing. (Leonard Lee Rue 111. Bruce Coleman Ltd.)

Fig. 85. Purple gallinule, *Porphyrula martinica,* this species damages rice by trampling and eating growing stems. (J. A. Hancock. Bruce Coleman Ltd)

dry season as the stock of dry seed from the previous rains diminishes. Given that quelea is possibly the most numerous bird in the world, and that most of the population lives in very remote areas of bush, it is unlikely that control teams could ever equal a 50% annual mortality by mass eradication. Even if this were possible, control would, in simple terms, merely ensure that all the survivors had sufficient food for the remainder of the year, in effect culling the population. The mortality imposed by man is not additive to natural mortality, but substitutes for it. In addition quelea performs regular migrations within Africa, so that reinvasion will occur regularly every year from areas outside those controlled by eradication teams. Thus in South Africa, where there exist excellent communications and efficient control teams many millions of queleas have been killed annually for over twenty years, yet there has been no long term reduction of the quelea population. The same considerations apply equally to other bird pest species; all are highly mobile, and it is now known that most of the African pest species perform regular annual migrations, so that reinvasion will recur every year.

While not offering any hope of long-term reduction of bird damage, such mass eradication methods may offer short-term relief. They may bring about a temporary reduction in the local bird population for long enough to allow grain to mature and be harvested without serious loss. In some situations in Africa damage to maturing wet-season rice by queleas and village weavers is caused by juvenile birds that have recently become independent from the breeding colonies. It is known that juvenile queleas do not move far from the colony for some weeks after fledging, so that destruction of all the breeding sites posing a potential threat to local rice crops may enable the harvest to be completed before young birds born elsewhere reinvade. Under these conditions quelea colonies remote from crop areas do not pose a threat and can be ignored, no matter how large they may be, with a considerable saving of expense and effort. Rice grown under irrigation during the dry season may be threatened by flocks of non-breeding birds. Again, depending on the rate of interchange of individual birds between roosts and the rate of reinvasion from other areas, destruction of selected roosts may achieve the desired relief from damage without it being necessary to embark on costly control of all bird concentrations over a vast area. In some situations the control effort required for 'immediate crop protection' is little less than that previously expended on total eradication, for example where valuable rice lands are so extensive that all bird concentrations may potentially cause damage (though even here control should be carried out only when there are vulnerable crops 'at risk').

An early but effective method of quelea eradication involved the placing by day of 44-gallon drums of petrol or diesel, together with a small gelignite charge, beneath the nesting bushes or amongst the trees where a roost had been located. As the birds returned at dusk they were driven by beaters towards the prepared site, and after dark the area was detonated. Often millions of birds were destroyed in a single spectacular operation (Fig. 87). However even in the dry season this method was not applicable to many areas where the transportation of heavy materials was difficult; during the rains it would frequently be impossible. Nesting sites were often destroyed much more simply by a small number of men armed with flame-throwers (Fig. 88). As with the explosives method the danger of starting bush fires is considerable. In recent years these methods have been replaced by the use of fixed-wing aircraft and helicopters for the aerial application of avicides such as parathion and fenthion. Aerial spraying is particularly useful for sites with difficult access from the ground, provided that the site can be adequately marked by scouts beforehand. This method is costly in terms of aircraft hire and pesticide, though it is probable that the total cost can be reduced by the use of low spray concentrations, yet still achieving a high mortality. The method is however extremely dangerous. In the case of roosting sites in particular the flying can only be carried out after dark at very low altitude. The chemicals used, being highly toxic to birds, are also highly toxic to other warmblooded animals such as livestock and man, and are also particularly toxic to fish. Parathion is now banned as an avicide in many countries, but even the relatively safer fenthion cannot be used in or near inundated areas where rice is grown, nor near areas of human habitation. These limitations apply particularly to the control of the village weaver, which, as its name implies, lives in very close proximity to man. In the United States, because of the dangers of environmental pollution, trials have been made in blackbird roosts using detergent solution as a spray. By overcoming the natural water-repellency of the plumage the birds are soaked, and provided the ensuing night is cold, they are unable to maintain body warmth and die. While this method is very effective on cold nights in temperate regions, the nights in rice growing regions are probably never cold enough, at least not during the growing season when control is required.

There are many situations where mass bird destruction is not possible, e.g. where it would pose a serious hazard to livestock and humans, or where the habits of the bird do not allow it, such as the golden sparrow which nests in extremely diffuse colonies. Eradication may be possible but undesirable for other reasons, as in the case of the black-tailed godwit, which caused considerable damage to newly planted rice seedlings in Senegal, yet is protected in Europe where it does no damage. In these situations strategy 2 has been adopted, and many methods have been devised to deflect bird damage, usually by the use of scaring devices. However, whatever the method used, whether it be physical or acoustic scaring, the use of repellent chemicals, or 'bird-resistant' varieties, the basic strategy requires that the birds be forced to return to their natural diet, or move elsewhere, or die. Yet for quelea at least, severe crop damage is almost certainly the result of the birds being unable to find their preferred natural food, and it has been frequently found that starving birds are impossible to dissuade. Against quelea in Tanzania all types of scaring devices proved equally ineffective after a few initial successes. In Senegal sirens, fumigation, and low-flying aircraft were ineffective. Some small success was achieved using a 37 mm gun mounted on a vehicle, but the results were not worth the effort required. Probably in such situations the birds merely take longer to consume the same amount. Where damage has been due to trampling, not to feeding, scaring methods may be successful. In Senegal luminous scarecrows erected in the rice fields at night and accompanied by detonations effectively prevented the use of the fields as roosts, and reduced trampling damage by wildfowl. Only 10% of the area had to be reseeded after this treatment compared with 30% previously. However, usually where success has been claimed it has been measured against damage in neighbouring fields that were not so protected. Thus any differences in crop damage

Fig. 86. Quelea nests in a typical colony. Note the enormous numbers of small round nests almost covering the trees. (FAO)

Fig. 87. Destruction of a quelea roost using high explosives and petrol, taken one second after the explosion. Up to five million birds have been roasted in a single explosion such as this. (P. Ward)

Fig. 88. Destroying small colonies of quelea nests using flame throwers. (FAO)

between protected and unprotected fields are probably exaggerated, with the unprotected fields suffering additional damage from the birds deflected from the protected ones. The total grain production of the region is probably little improved, if at all. Thus any reduction of damage by the adoption of a new deterrent method is probably at the expense of neighbouring farmers, who must quickly do likewise to restore the status quo. Then, faced with the choice of starvation or taking 'protected' grain, the only type available, they continue to do damage as before. The only value of disuasive methods may be simply to share the burden of crop damage equally between all the growers in an area, a function already achieved cheaply and successfully for centuries by traditional bird-scaring methods.

Ideally a category 3 strategy should be adopted wherever possible; the avoidance of damage by a change in agricultural practice. For example since 1972 damage to seedling rice in Senegal has been avoided by delaying planting, which previously took place from April onwards, until the end of July when palaearctic wildfowl have departed on migration, and African species are involved in breeding activities over a wide area and do not congregate on the rice fields. Damage still occurs however from October onwards when the palaearctic wildfowl return and local

216

species again congregate in flocks. Areas of rice where growth has been poor and the plants are short and sparse are then attractive to ducks and damage may occur. Careful cultivation and weeding to produce a field of uniform good growth unattractive to wildfowl, coupled with conservation of marginal marshy areas specifically for wildfowl may considerably reduce trampling damage. Where rice growing is carried out under irrigation it may be possible to time the growing season such that harvest can be completed before it becomes vulnerable to serious bird damage. In Africa granivorous birds must undertake a regular migration each year out of their dry season quarters at the start of the rains as their natural seed food germinates and becomes unavailable. The birds cannot return until some two months later when new seed becomes available, and do not produce young, which do most of the damage, for a further month. Farmers in NW Botswana who practice early planting of millet and sorghum, and harvest before queleas return to the area, or at least before March when quelea young leave the colonies, suffer little damage. The rice growing potential of the region is considerable, but damage by quelea will also be severe unless early planting, or quick maturation and early harvest are an integral part of crop management.

In conclusion, whatever the method of damage reduction adopted it will fall into one of the three categories outlined above. If that strategy is inapplicable to the particular pest situation then the control method used, however sophisticated, will fail.

Bibliography

CARRICK, R. (1956). The Little Corella, *Kakatoe sanguinea G.*, and rice cultivation in the Kimberley region W.A. *CSIRO Wildlife Research* 1: 69–71.

HAYLOCK, J. W., DISNEY, H. J. de S. and RAPLEY, R. E. (1954). Control of the Sudan Dioch or Red-billed Finch in Tanganyika. *East African Agricultural Journal* 21: 210–217.

JONES, P. J. (1975). The significance of bird migration to bird pest control strategy. *Proceedings 8th British Insecticide Fungicide Conference* Vol 3.

LONG, J. L. (1969). The Java Sparrow. *Journal of Agriculture of Western Australia* 10 (5): 212–213.

MOREL, G. (1962). Quelques methodes d'effarouchement des oiseaux utilisées a Richard-Toll (Sénégal). *Annales des Epiphyties* 13 (special issue) pp. 258.

ORIANS, G. H. (1960). Autumnal breeding in the Tricolored Blackbird. *Auk* 77: 379–398.

POPE, G. G. and WARD, P. (1972). The effects of small applications of an organophosphorous poison, fenthion on the weaver bird *Quelea quelea*. *Pesticide Science* 3: 197–205.

TRECA, B. (1975). Les oiseaux d'eau et la riziculture dans la delta du Sénégal. *Oiseaux* 45: 259–265.

WARD, P. (1971). The migration patterns of *Quelea quelea* in Africa. *Ibis* 113: 275–297.

WARD, P. (1973). A new strategy for the control of damage by Quelas. *PANS* 19 (1): 97–106.

WILSON-JONES, K. (1962). Rice growing in N. Australia. *Tropical Science* 4: 181–204.

RODENTS

Introduction

Rodents attack rice at all stages of growth from planting to harvest and, if given the opportunity, will continue to attack the grain whilst it is in store. Freshly sown rice may be dug up and the seed eaten or the soft growing tissue of seedling rice may be gnawed. As the seedlings grow into young rice plants, rodents turn their attention to the heart of the stem, often discarding the leaves. The grain is eaten by rodents from the time it starts to form until harvest, each stem generally being felled by gnawing at 5–15 cm above ground level. Some species may store quantities of grain in their burrows. Large rodents, beside feeding on the crop, may cause serious damage to the bunds or levees. Little is known about extent of losses caused by rodents, since where damage occurs, its seriousness is usually apparent and effort is put into control rather than into measuring the amount of the crop that has been lost. Total loss due to rodent attack in a ricefield can occur but is unusual; more often, a proportion of the crop is lost in each field, with damage sometimes extending over large areas. Particularly severe rodent damage to rice has occurred in the Philippines from time to time. In 1958 for instance, losses on the island of Mindanao were so severe that the Government had to send relief food supplies to prevent starvation. Similar widespread damage by rodents also occurred during the early stages of the Wageningen Rice Project in Surinam where, in 1957, 250 ha of the 1,000 ha planted were entirely destroyed.

Rodent damage to rice has been reported from many other parts of the world including Indonesia, Spain, India, Guyana and California. Most reports give little information about the nature or extent of damage: the authors generally appear to assume that rodent damage to rice is commonplace, needing no explanation, and it may be fair to assume that rodent damage occurs to some extent in ricefields in all areas where the crop is grown.

The Food and Agriculture Organisation of the United Nations has estimated that throughout the world rodents damage or destroy 12 million tons of rice annually with most of the loss occurring in store. But losses in the field are often greater than in store, and so the overall rodent problem is serious. In fact, the opinion has been expressed that, with the current use of pesticides against the invertebrates and fungi that attack rice, losses due to rodents and birds are probably greater now in many parts of the world than those caused by all other pests.

RODENT BIOLOGY

Species that cause damage

Many species of rodents attack rice and frequently the principal mammal pest in any area is an indigenous rat-like rodent that probably formerly inhabited grassland or swamp. Species that normally live in buildings (i.e. commensal rodents)

TABLE 18. SOME RODENTS THAT ATTACK GROWING RICE

Region	Scientific and common name	Remarks
Philippines	*Rattus argentiventer* ricefield rat	Most important rice pest on Mindanao and Mindoro (previously known as *Rattus rattus umbriventer*).
	Rattus rattus mindanensis Philippine house rat	Most important rice pest on Luzon and the Visayas.
	Rattus exulans Pacific island rat	Of minor importance as a rice pest.
Indonesia	*Rattus argentiventer* sawah rat	The major rice pest throughout Indonesia (previously known as *Rattus rattus brevicaudatus*).
West Malaysia	*Rattus argentiventer* ricefield rat	An important and widespread rice pest.
	Rattus tiomanicus Malaysian wood rat	Important in parts of north Malaya.
Australia	*Mus musculus* house mouse	Severe damage to rice during periodic mouse outbreaks.
Thailand	*Rattus argentiventer* ricefield rat	Widespread major pest of rice.
	Bandicota indica greater bandicoot rat	Widespread.
	Rattus losea	Important in North Thailand.
	Rattus exulans Pacific island rat	
Taiwan	*Rattus losea*	
	Bandicota nemorivaga	
Indian sub-continent	*Bandicota bengalensis* lesser bandicoot rat	Notable rice pest which burrows extensively and stores quantities of rice underground (Figs. 89–91).
	Nesokia indica short-tailed bandicoot rat	
	Millardia meltada grass rat	
	Tatera indica Indian gerbil	Confined to drier areas.
	Mus spp. mice	Occasionally very numerous, causing more damage than the much larger rats.
Europe	*Rattus norvegicus* Norway rat	Spreads from villages and farms to ricefields.

TABLE 18—continued

Region	Scientific and common name	Remarks
Africa	*Mastomys natalensis* multimammate rat	Widespread and important pest; numbers tend to fluctuate (Fig. 92).
	Arvicanthis niloticus grass rat	Widespread; numbers fluctuate.
	Thryonomys swinderianus cane rat or cutting-grass	A large rodent only locally important as a pest (Fig. 93).
North America	*Rattus norvegicus* Norway rat	
	Rattus rattus roof rat	
	Myocaster coypus coypu	An introduced species important in Texas (Fig. 94).
	Ondatra zebithica muskrat	Damages bunds.
Central America	*Sigmodon hispidus* cotton rat	Numbers fluctuate; severe damage at times.
	Oryzomys spp. rice rats	
South America	*Holochilus sciureus*	Semi-aquatic species causing severe damage in Surinam and elsewhere.
	Oryzomys spp. rice rats	

also invade ricefields in some areas and a few large rodents such as the coypu sometimes cause damage. Some of the better known rice pests are shown in Table 18.

The proper identification of a field rodent pest is important, although this cannot always be undertaken on the spot. Many rat species appear outwardly similar but exhibit different behaviour and may respond differently to control measures used against them. In some regions taxonomic keys exist for the identification of rodents but without experience, these are not always readily followed. In any case, on-the-spot identification should be checked by a reputable local museum. Instructions on preparing specimens for museum identification have been published by the British Museum.

Life history and behaviour

Reproduction

Many of the rodents that are serious pests of rice are rats but the reproductive potential of rat species differ. For instance, *Holochilus* in Guyana may have a

Fig. 89. Developing earheads of rice severed by the feeding of *Bandicota bengalensis*. (H. Fernando)

Fig. 90. Platform made by *Bandicota* feeding on rice at the booting stage. (H. Fernando)

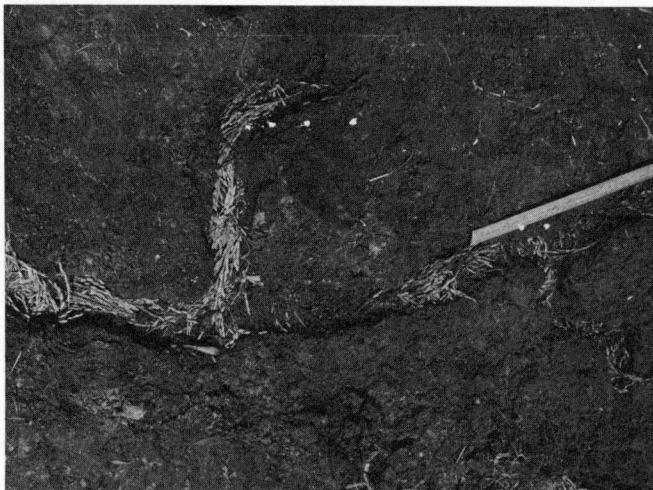

Fig. 91. Section of a ricefield bund showing rice earheads stored in *Bandicota* burrows. (H. Fernando)

litter size of only 3—4 whereas commensal *Rattus* species may average 6—8 young per litter and the small African *Mastomys natalensis* may average as many as 11—12. However, reproductive rate is also dependent on other factors such as the length of the breeding season, the age at which females become mature and the number of litters that a female can produce in a given time. Thus, although commensal rats often breed throughout the year and have a gestation period of only 21—24 days, individual females may still produce as few as five litters per annum due to the occurrence of 'resting periods' during which no litters are produced.

Little attention has been paid to the reproductive cycles of many of the rats that attack rice, but it is known that in the Philippines the principal pest *Rattus argentiventer* reproduces rapidly during the seven months of the year that rice is grown, but almost ceases to breed during the remaining dry months of January to May. In some areas the species may therefore be largely dependent on rice. A relationship between reproduction and the wet season also occurs in the multimammate rat and in Guyana, young *Holochilus* may start to reproduce at an earlier age during the wet season than during the dry season. The sawah rat (ricefield rat) of Indonesia breeds only during the ripening and harvest of rice, when it normally produces three litters in rapid succession.

It is clear that rats have a potential for rapid increase in numbers when conditions are suitable for reproduction, and for this reason rats are usually the most serious pests in seasonal crops such as rice; the larger, slower-breeding rodents are usually not able to increase their numbers rapidly enough during the course of a season to become serious pests.

Food

Rodents that in their normal habitat feed on wild grass seed and the seed of other plants readily turn their attention to cultivated grain. But, of course, ripe grain is only present for relatively short periods of the crop cycle, namely for a few days after sowing and again shortly before until shortly after reaping. During the growth period of the rice crop therefore, granivorous rodents may be forced to feed on the rice stems. Frequently, only a small section of the heart of each stem is eaten, the outer sheath and leaves being discarded. A single field rat can ruin as many as 100 stems of 4–6 weeks old rice per day. This type of damage is therefore often more serious than damage to the maturing panicles, each of which may provide a full meal for a rat.

The rice stem is clearly of poor food value to at least one species of ricefield rat. In tests, the sawah rat of Indonesia, *Rattus argentiventer*, survived for as little as 5 days when fed on young rice stems and for 9 days on a diet of flowering rice. The more omnivorous commensal rodents also do not survive well and cannot reproduce on a diet of plant material, as was reported for rats fed on sugarcane. It is therefore probable that many of the rat species that attack rice, although they feed heavily on the stems, require other protein- or carbohydrate-rich food sources in order to survive and reproduce during the growth period of the crop. Such sources of food for rats are provided in ricefields by weeds both within the

Fig. 92. Multimammate rat *Mastomys natalensis* a serious pest throughout Africa. (Jane Burton)

crop and on the surrounding bunds, by insects, crustacea and molluscs and sometimes by other cereal crops grown in proximity to the rice.

Movement

The relatively short duration of rice and other cereal crops results in an unstable habitat for rats and other animals. Before harvest there is a food surplus and rodent populations tend to reach their peak, but afterwards the amount of food available decreases very rapidly and rodents must either die or move elsewhere in search of food. In Indonesia, field rats become numerous in the villages shortly after the rice harvest. Patches of bush or swampland neighbouring ricefields are probably infested by field rats at all times and form important refuges from which rats can reinfest the crop. Movement appears generally to be a gradual process although there are a few reports of rats moving *en masse*.

Rats are undoubtedly capable of travelling long distances in search of alternative food sources. Distances of 400–700 m have been recorded for species infesting ricefields. But rats on the move are generally more vulnerable to predation than animals with fixed home sites and established feeding places and many of the rats that emigrate from ricefields after harvest probably perish. The proximity and extent of suitable alternative habitat is very important in determining the numbers of immigrating rats that survive to form the nucleus from which succeeding crops of rice are infested.

Under normal circumstances, rats tend to move minimum distances from their home sites to obtain food. In the Philippines the ricefield rat generally travels only 20 to 40 m.

Diseases carried by rodents

Field rodents as well as the commensal species can carry diseases that affect man. Thus bubonic plague is endemic amongst field rodents in some parts of the world and is only transmitted to man when the local house rats become infested. However, ricefields do not represent any particular plague risk since the insect vectors, fleas, do not survive well in the damp conditions that often prevail in rice. On the other hand, irrigation favours the transmission of other rodent-borne diseases, such as salmonellosis and particularly leptospirosis.

Leptospirosis, one form of which is known as Weil's disease, is caused by a motile organism voided in the urine of rodents and other animals. In dry conditions leptospires quickly die, but in water, particularly if it is slightly alkaline, they may survive for several days. Ricefields may therefore form ideal conditions for the transmission of the disease to man. The organisms enter the human body through skin abrasions or through the mucous membranes of the nose and throat. The disease has several clinical forms, some of which are comparatively mild, and it is often unrecognised unless doctors and hospitals are alerted to its presence in the area.

Prophylaxis against the disease has been attempted in three ways; by killing the rats that harbour the disease, by killing the leptospires in the paddy water and by immunising ricefield workers. Of these methods, immunisation has probably been used most widely to greatest effect. However, rat control as a prophylactic measure also has the advantage of preventing damage to the rice. Killing leptospires in the paddy water has been recommended in Japan by using the fertilizer, calcium cyanamide, which is a more effective leptospirocide than other commonly used rice fertilizers.

Signs of infestation and circumstances of attack

In many areas where ricefields are attacked by rats, there is a tendency for a strip, about 1 m wide around the edges of fields to remain undamaged. It is therefore often necessary to penetrate the crop a little way before the extent of damage can be observed (Fig. 95). In flooded paddyfields, feeding by rats tends to take place on raised, dry areas within the fields but *Holochilus* is South America also feeds freely in the flooded areas, building circular platforms of cut rice stems on which it may nest. Evidence of the presence of rats is also provided by their holes and runways into and along the banks separating paddyfields.

Severe rodent damage to rice often occurs in areas where rice is a crop of secondary importance and ricefields are interspersed with other crops attractive to rodents. Thus a relatively dense rodent population may be maintained by a variety of foods and the rice attacked as soon as it reaches the susceptible stage. Weedy fields and fields close to uncultivated land or unweeded bunds are also more liable to rodent damage than fields in areas where 'clean' cultivation is practised.

Fig. 93. Cutting-grass, *Thryonomys swinderianus* a pest in African rice-growing areas. (Jane Burton)

Fig. 94. Rice damaged by coypu in Texas. (Roger Nass)

Fig. 95. Centre of a ricefield showing rice ripening late after rat attack. (Note rice at margins of field nearly mature). (H. Fernando, Central Agricultural Research Institute, Sri Lanka)

Outbreaks of rats that devastate crops over a wide area are comparatively rare on settled rice-growing land and their causes are by no means fully understood. However, outbreaks seem to be particularly likely in recently cultivated areas. For instance the outbreak that occurred in Cotabato Province of Mindanao (Philippines) in 1953 followed the settlement of farmers in an area that had been sparsely cultivated previously. The first crops from the Wageningen rice project in Surinam also suffered from a severe rat outbreak. It appears that the debris left from the clearance of bush land and the general untidiness following clearance, provide ideal conditions for the build-up of certain rat species. Unusually heavy or prolonged rainfall, by stimulating weed growth and hampering the harvesting of cereals, may also provide conditions for a rat outbreak.

CONTROL

Keeping out rodents

Land management

Lasting control cannot be achieved by killing rodents — reinfestation always occurs sooner or later through reproduction amongst the survivors or through immigration. Of course, methods of killing rodents can be used repeatedly, but there is a tendency for some methods — particularly poisoning — to become less effective with continual use; furthermore, the cost of supporting rodent control indefinitely is considerable. Methods of killing rodents should therefore be regarded merely as interim measures.

There are several important ways in which damage to rice can be reduced without killing the rodents concerned. Weed control, both within the crop and along the bunds and dykes separating fields has an important limiting effect on rat population. The clearance of bush or swamp close to ricefields also limits the number of rodents by reducing sources of food and shelter during periods of the year when no rice is available. Synchronised planting of rice over wide areas reduces the period during which the crop is available to rodents so that they have to rely on other food sources for most of the year. Further reduction in alternative food sources for rodents can be achieved by encouraging a monoculture system in areas particularly suitable for rice. Maintaining the water level in ricefields may prevent rodent damage to germinating seed and reduce damage at other times. The timing of the harvest is also important. Harvesting before the rice is fully ripe both saves grain that might be lost by shattering and, in so doing, allows less grain to fall to the ground for rodents to pick up afterwards.

Proofing

Exclusion of rodents is the most effective method of preventing damage to stored products but is not practicable on a large scale in the field. However, prevention of rodent damage to experimental plots of rice is highly important and a method of excluding rats from ricefields using an electrified fence has been developed in the Philippines. Two wires are supported on insulators at 1–2 cm above 30 cm high wire-netting fence, running around the field. Current from a 6 or 12 volt battery is transformed to 125 or 250 volts and fed into the wires so that rats climbing the fence are electrocuted. The system is said to be potentially dangerous to larger animals including man, and the current is generally switched on only at night. Regular patrolling is necessary to prevent dead rodents earthing the live wires.

Killing rodents

Choice of method

Rodents vary in their reactions to control measures and may alter their behaviour in response to attacks made against them. Control procedures for each pest species therefore cannot be standardised. Instead, a control method should be chosen to suit each particular set of circumstances; furthermore, assessment of the effect of the method should be made an integral part of the procedure so that any shortcomings can be detected early and a more appropriate method substituted. Thus rodent control is often a matter of trial and error with success depending less on what method is used than on the alertness of those conducting the treatment.

Poisons

There are two main types of poison used gainst rodents; these are known as acute and chronic poisons. Acute poisons kill after a single dose and the onset of symptoms is rapid. Rapid action is a practical disadvantage in the use of these poisons since an animal has only a limited period in which to consume a lethal dose before symptoms arise and feeding ceases. Furthermore, animals that ingest sub-lethal quantities of acute poisons usually avoid the bait or poison for a period of weeks or even months afterwards. This behaviour is known as bait or poison shyness. Methods of using acute poisons are therefore largely concerned with inducing rodents to feed rapidly on the poisoned bait.

The chronic poisons that are commonly used against rodents are all anticoagulants — that is, they reduce the ability of the blood to clot, so that affected animals die of haemorrhage. Anticoagulants are used at comparatively low concentrations in bait so that a rodent must feed several times before ingesting a lethal dose. The onset of symptoms is slow and animals do not associate illness with feeding on the bait, and may even continue to feed after consuming lethal quantities of the poison. Animals vary in their susceptibility to anticoagulants and in a few limited areas of Europe, resistant colonies of the Norway rat (*Rattus norvegicus*) have developed following repeated use of these poisons. Anticoagulant resistance is, however, a rare occurrence and ricefield rats can safely be regarded as susceptible until evidence is found to the contrary.

Using acute poisons

A list of the more useful acute poisons together with concentrations (in cereal bait) at which they should be used against known susceptible species is given in Table 19. Zinc phosphide is probably the most generally useful although it may not be as effective as the more toxic compounds such as sodium fluoroacetate,

TABLE 19. SOME ACUTE POISONS USED AS RODENTICIDES

Poison	Per cent by weight in cereal bait
Zinc phosphide	1—5
Thallium sulphate	0.5—1.5
Fluoroacetamide	2
Sodium fluoroacetate (1080)	0.25
Norbormide (Raticate)	0.5—1

fluoroacetamide and thallium sulphate which, because of their extremely hazardous nature can only be used under the most rigorous safety conditions. Norbormide, on the other hand, is comparatively harmless to animals other than rats of the genus *Rattus*, but is less effective than most acute rodenticides.

Acute poisons are generally used mixed with a cereal bait which can be either milled, broken or whole grain; in the latter two cases, a more even mixture is obtained if oil or water is added to cause the poison to adhere to the bait. Best results with acute poisons are obtained if plain (unpoisoned) bait is laid from 4 to 8 days before being replaced with poisoned bait. This technique, known as pre-baiting, allows rodents to get used to feeding at a particular place and on a particular bait, so that when the poison is laid they will feed on it readily and rapidly.

Acute poisons are most effective if laid, after pre-baiting, close to the places where rodents are living. However, the labour involved in treating large areas is considerable and if less effective control is acceptable, means of broadcasting the bait either manually, by grain seeders or from the air can be used.

Using chronic poisons

There are several anticoagulants available as rodenticides of which warfarin is the best known. They are used at concentrations which make them more or less comparable in toxicity and any differences in effectiveness between them are therefore most likely to be due to differences in acceptability. Since, however, the latter is not only affected by the poison and its concentration, but also by additives incorporated by the manufacturer and by the bait in which it is finally mixed, satisfactory advice on the choice of an anticoagulant can only be given after investigation in the area concerned.

Many anticoagulants are marketed already mixed (at concentrations from 0.005 to 0.05% in a milled cereal bait. Ready-to-use poison bait like this is often expensive, and has the disadvantage of preventing variation in the bait to suit different circumstances. Thus, it is generally best to use concentrates (master-mixes), which are marketed in various strengths — usually in the range, 0.1 to 1.0% active ingredient.

As with acute poisons, cereal baits are generally most useful and if coarse ground, require a 'sticker' to cause the poison to adhere to the bait. Mineral oil may be used for this purpose but vegetable oils may initially increase the attractiveness of the bait, although they may later turn rancid. Another useful attractant is sugar which should be mixed at about 5%. Ideally, local materials should be tried in the field to determine which can be used to greatest effect. In damp and humid conditions mould formation can be inhibited by the addition of 0.25% paranitrophenol or 0.1% dehydroacetic acid. However, both these additions may lower the palatability of the bait and should only be used when absolutely necessary.

Pre-baiting with anticoagulants is entirely unnecessary and is wasteful of time, labour and bait. Piles of anticoagulant bait should be laid directly in places where rodents will find and eat them before reaching their normal food. Thus bait should be laid close to holes, runways or places where rodents are known to be. In wet ricefields this generally means laying the bait at selected points along the bunds between fields. Bait points should be as close together as is practicable — not more than 15 m apart on a continuously infested bund. The amount of bait in each pile should be initially about 100—200 g and thereafter a surplus maintained at all points by revisiting them every few days and laying more bait where necessary. Some form of cover must be provided for each bait point both to protect it from the weather and to exclude other animals as far as possible. Lengths of large diameter bamboo or piping are used extensively for this purpose (Fig. 96) but are not altogether satisfactory since heavy rain often splashes into them, dampening the bait. Metal containers protected by rectangular pieces of sheet metal, curved to form an arch are more satisfactory but, of course, are also

Fig. 96. Rat bait laid in large diameter bamboos along the edge of a trial plot at the International Rice Research Institute. (Note electric rat fence). (Susan D. Feakin)

more expensive. Small thatched shelters are built to protect bait in parts of the Philippines.

Baits can also be protected by wrapping them in envelopes of waterproof paper or polythene, but this allows moisture to enter as soon as the envelope is penetrated by the rodent and the bait may be spoiled before the rodent has had time to consume lethal quantities of it. Another and perhaps more promising approach is the formulation of bait blocks using paraffin wax or stearic acid. These are easily made. Eleven parts of cornmeal, 2 parts of sugar and 1¼ parts of 0.5% warfarin powder are thoroughly mixed. Eight parts of paraffin wax are melted in a suitable container. The dry mixture is stirred in as soon as the wax has melted and the mixture immediately spread out on trays 2.5 cm deep, allowed to set and then cut into blocks of about 227 g. Warfarin is thermolabile and it is important that it is not added to the wax at too high a temperature. When made with care these blocks will have a warfarin content of not less than 0.025%. These blocks have been successfully used against ricefield rats in Malaysia.

After about a week of using anticoagulants the amount of bait eaten should be rapidly declining as animals die and ultimately feeding on the bait should cease altogether. Cessation of feeding should occur roughly within 3 to 4 weeks — although much depends on the circumstances of the treatment. If, after feeding on the bait has ceased, evidence is found that some rodents have not been killed, a different bait should be tried and the number of bait points increased at the appropriate places. In some circumstances, bait may continue to be taken for long periods and this may happen because there is continuous reinfestation. Evidence of reinfestation will generally be available in the form of fluctuations in the distribution and amount of rodent activity and of recently dead animals continuing to be found on the site. This commonly occurs in areas of small farms where only a proportion of the farmers are doing rodent control. The proper approach to this situation is to encourage all farmers in the area to act simultaneously (Fig. 97), but where this fails, those farmers that are interested can often maintain a reasonable degree of rodent control by continuing the anticoagulant treatment for as long as the crop is susceptible.

Other methods

The poisons already dealt with are effective through being eaten by rodents; there are others that are absorbed through the lungs. Powders that produce hydrogen cyanide gas when they come into contact with moisture can be used to control rodents in burrows made in well consolidated damp soil where a good concentration of gas can be built up. A teaspoon of gassing powder should be placed well down each hole and the hole blocked with earth. Burrows found to have been reopened by rats the following day should be re-treated and blocked again, and the procedure repeated until holes cease to be reopened. Good results can also be obtained using aluminium phosphide tablets instead of cyanide powder. Gassing powders are sometimes pumped into rat burrows but this method is of doubtful value as many of the rats tend to bolt.

Trapping is rarely completely effective since rodents readily become trap-shy, but it can be useful in eliminating a few survivors from a poisoning campaign. 'Breakback' traps are inexpensive and effective but local designs may be preferred such

Fig. 97. Dead rats recovered after a night's poisoning organised by the Bureau of Plant Industry, Manila, Philippines. (Bureau of Plant Industry, Manila)

Fig. 98. Rat control in ricefields in Andhra Pradesh, *Bandicota bengalensis* caught in a Tanjore bamboo trap. (Ministry of Food and Agriculture, India)

as the Tanjore Bamboo trap (Fig. 98). Trapping, as with poison baiting is more effective when natural rodent food is scarce, and in ricefields may fail completely after the grain forms.

Rodents have been killed by a 'blanket' system whereby villagers drive them from cover in grassland to a prepared mound of cut grass which is then taken apart gradually and the animals killed with sticks. Heavy rollers pulled by tractor have also been used to crush rodents in their burrows along ricefield bunds. However, these specialised methods are not generally applicable.

Prevention of infestation

The rodents that infest ricefields generally multiply during the crop season reaching peak population at or soon after harvest. Thereafter a rapid decline may set in, with only small colonies surviving here and there until rice is next planted. Control can be conducted more effectively while rodents are confined to relatively small areas where food is scarce than when they are widely dispersed amongst the crop. The maintenance of permanent anticoagulant baits may be all that is necessary to control rats in small patches of bush or grassland, but larger areas may need to be treated with acute poisons, perhaps using mechanical means to disperse the bait. However, before preventive measures of this type can be undertaken with expectation of success, it is necessary to understand something of the behaviour and ecology of the pest species in the area concerned. Detailed research will provide many of the answers but this should be backed up by field trials of various methods in which adequate provision is made for experimental controls.

Bibliography

ABRAHAM, E. V. (1958). Bothered by rats in rice fields? *Indian Farming* 7: 31–32.

ALFONSO, P. J., MORALES, J. L. and SUMANGIL, J. P. (1965). Control of ricefield rats, in *Rice Production Manual,* pp. 237–252. University of Philippines College of Agriculture.

ALTAVA, D. and BARRERA, M. (1961). La desratizacion como profilaxis de la leptospirosis de los arrozales. *Revista de Sanidad e Higiene Pública* 35: 647–655.

BABUDIERI, B. (1953). Epidemiology of leptospirosis in Italian ricefields. In *Advances in the control of zoonoses.* World Health Organisation, Geneva. Monograph series No. 19.

BABUDIERI, B. (1960). Vaccine and vaccination against human leptospirosis. Polish Academy of Sciences, 19th International Symposium. *Leptospirae and leptospirosis in man and animals.* Lublin, 1958.

BREWER, W. E., ALEXANDER, A. D., HAKIOGLU, F. and EVANS, L. B. (1960). Ricefield leptospirosis in Turkey. A serological survey, *American Journal of Tropical Medicine and Hygiene* 9: 229–239.

BRITISH MUSEUM (1968). Instructions for collectors No. 1: mammals. British Museum (Natural History). pp. 55

CAUM, E. L. (1922). Why do rats eat cane? *Hawaiian Planters Record* 26: 213–215.

CLARK, R. J. (1958). The control of field rats in Mindanao. FAO Report No. 785, pp. 29.

CLOWES, H. G. (1950). Rice production and utilisation. *Nature* 166: 168–170.

DOBROVSKY, T. M. (1966). Field rats. FAO *Grain Storage Newsletter* 8 (3): 65–66.

DOTY, R. E. (1945). Rat control on Hawaiian sugarcane plantations. *Hawaiian Planters Record* 49: 71—239.

FERNANDO, H. E., KAWAMOTO, N. and PERERA, N. (1967). The biology and control of the ricefield mole rat of Ceylon, *Gunomys gracilis*. *FAO Plant Protection Bulletin* 15: 32—37.

GOOT, P. VAN DER (1951). Over levenswijze en bestriding van sawah-ratten in het laagland van Java. *Landbouw* (Bogor, Java) 23: 123—284.

KENNARD, C. P. (1965). Pests and diseases of rice in British Guiana and their control. *FAO Plant Protection Bulletin* 13: 73—78.

MARSH, R. E. (1966). Methods of controlling rodents and birds in ricefields. *Congres de la Protection des Cultures Tropicales*, Marseille. pp. 633—637.

OLIFF, W. D. (1953). The mortality, fecunditiy and intrinsic rate of natural inrease of the multimammate mouse in the laboratory. *Journal of Animal Ecology* 22: 217—226.

RAMOS, F. V. (1967). The electric rat fence. *International Rice Research Institute, Los Banos, Philippines, Technical Paper* No. 3, pp. 4.

REIFF, H., BUXTRORF, A. and VILA, J. P. (1954). Un essai de lutte contre les rats gris dans les rizieres de Valence, en une de la prophylaxie de la leptaspirose. Information, J. R. Geigy S.A., Bâle, 25 pp.

RYU, E. and LIU, C. K. (1967). Studies on leptospirocidal action of fertilisers. *Journal Taiwan Association of Animal Husbandry and Veterinary Medicine* 11: 43—47.

SMITH, C. E. G. and TURNER, L. H. (1961). The effect of pH on the survival of Leptospires in water. *Bulletin World Health Organisation* 24: 35—43.

SMITH, R. W. (1967). The control of rats in coconuts using rat blocks. *Oleagineux* 22: 159—160.

SUMANGIL, J. P. (1965). *Handbook on the ecology of ricefield rats and their control by chemical methods.* Bureau of Plant Industry, Manila, pp. 27.

TAYLOR, K. D. (1968). An outbreak of rats in agricultural areas of Kenya in 1962. *East African Agricultural and Forestry Journal* 34: 66—77.

TOWNES, H. and MORALES, J. (1954). Control of field rats in the Philippines with special reference to Cotabato. *Proceedings 8th Pacific Science Congress, Quezon City*, 1953.

TOHYAME, Y. (1927). Result of prophylaxis of Weil's disease experimented in Kagoshima Prefecture. *Science Report Institute of Infectious Diseases, Tokyo University* 6: 555—557.

TWIGG, G. I. (1965). Studies on *Holochilus sciureus berbicensis* a cricetine rodent from the coastal region of British Guiana. *Proceedings Zoological Society of London* 145: 263—283.

WIT, T. P. M. de (1960). *The Wageningen rice project in Surinam.* Stichting voor de ontwikkeling van machinale landbouw in Suriname, The Hague, Netherlands. pp. 293.

WOOD, B. J. (1971). Investigations of rats in ricefields demonstrating an effective control method giving substantial yield increase. *PANS* 17: 180—193.

STORAGE

The many forms in which rice may be stored, and the variation in degree of susceptibility of those forms to the biological factors which are responsible for some types of deterioration during storage, makes this one of the most complex of all the cereal grains. Unlike the majority of cereals, rice is consumed as a whole grain, so that the presence of these organisms is of importance, not only because of the resulting loss in weight and quality but also because of their effect, particularly in the case of insects, upon the whole grain percentage.

Rice may be stored in the unhusked state, generally referred to as paddy or rough rice (raw or parboiled according to whether or not it has been subjected to the process of parboiling), after removal of the husk, as hulled rice or, after further processing, as milled rice (again raw or parboiled). Milled rice may be termed under-milled, medium-milled, fully-milled or polished according to the degree to which the outer layers (pericarp and aleurone layer) and the embryo have been removed and the grains surface-treated.

Parboiling is a process of steeping paddy in water for the period of one to three days and then subjecting it to steam at low pressure before drying and milling in the ordinary way. The result of parboiling is to make it easier to remove the husk and for this reason less broken rice is obtained and the amount of milling necessary is less than for the preparation of white rice. The parboiling as usually carried out in Asia gives the grain a yellowish colour and a distinctive flavour which render it unacceptable in many markets both in Asia and the Western Hemisphere. About 57% of the rice produced in India is parboiled, in Burma it accounts for 22% of the total rice produced; it is also popular in many parts of Africa and the West Indies.

Parboiled rice is more nutritious than white rice because it retains a higher proportion of vitamins and nutrients. It is less liable to insect attack and has better keeping qualities than white rice. When cooked, it has better qualities than white rice since it does not readily turn sour.

There are many factors, biological, chemical and physical which, singly or in combination, can lead to quality loss in rice, but it is relevant here to discuss only those aspects of loss, quantitative and qualitative, which result from attack by insects and moulds in storage and to consider the physical conditions conducive to such attack.

Storability of paddy and rice is affected by a number of pre-harvest factors, in particular moisture content. At the time of harvest, the moisture content of paddy may be as high as 30% and it is important, therefore, that drying (Fig. 99) should begin immediately. In countries where rice is harvested by hand there is sometimes a delay between harvesting and threshing (Fig. 100) and, often, the panicles are placed in heaps on platforms (or on the ground) in the ricefield, or in the village until they can be dealt with. Such temporary storage of grain with

Fig. 99. Sun-drying paddy after threshing. (Pest Infestation Control Laboratory)

Fig. 100. Heap of panicles in the field awaiting threshing. (Pest Infestation Control Laboratory)

238

a high moisture content will lead to the development of high temperatures and serious deterioration in the form of discolouration and off-odours, and ultimately to a reduction in milling yield, due partly to the growth of moulds and other microorganisms. This reduction may be avoided by arranging the panicles over racks in such a way that drying may commence through the combined effects of sunshine and air movement.

In order to minimise the degree of breakage of grains in the milling of raw paddy, careful attention must be paid to post-harvest drying because of the tendency of the grains to suffer internal cracking (or 'checking') if the paddy is allowed to be affected by a cycle of alternate wetting and drying. In the use of artificial drying particular attention should be paid to the need for multi-stage drying which allows for tempering periods during which equilibration of the grain may take place. The susceptibility of milled rice to attack by some insect species is increased according to the proportion of broken grains present.

For safe storage, and to obtain the best milling results, the moisture content of paddy should not exceed 14% (i.e. that moisture content which is in equilibrium with 70% relative humidity). The moisture content/relative humidity relationships of milled rice are to some extent dependent upon the degree of milling and whether or not parboiling has been carried out (e.g. the moisture content in equilibrium with 70% relative humidity may fall within the range 11.5—15%). In general it can be said that the moisture content of milled rice should never exceed 14% and that, in order to minimise attack by insects and prevent the growth of moulds it should, wherever possible, be reduced to 11.5%.

MOULDS

Reference has already been made to the deterioration which may result from the growth of moulds on stored paddy grains if these are not adequately dried. Similarly, if milled rice is stored under damp conditions, mould growth will occur. Apart from direct spoilage as a result of mould growth (see p. 237), the presence of moulds may encourage the development of some species of insects (e.g. *Alphitobius* spp., see p. 244). It is also known that the growth of moulds on paddy and rice can lead to the presence of toxic metabolites or mycotoxins. Reference has been made in *PANS* Manual No 2 (*Pest Control in Groundnuts*) to aflatoxin, which is produced by the mould *Aspergillus flavus* when this is allowed to grow in groundnuts and groundnut products. It can also infect rice. As well as aflatoxin several other mould toxins have been found in rice.

The microflora of a sample of paddy or rice will depend upon the method of cultivation, the climatic conditions at the time of harvest and subsequent drying and storage conditions. Both 'field' moulds such as *Helminthosporium oryzae*, *Epicoccum purpurescens*, *Gibberella fujikuroi* and *Pyricularia oryzae*, and, 'storage' moulds such as *Penicillium*, *Aspergillus* and *Rhizopus* species may occur. The optimum conditions for growth of most storage moulds are a temperature within the range $20°—40°C$ and a relative humidity above 70% (in many cases above 80%).

Tests of fungicides have not revealed any which can be relied upon to control those mould species troublesome on stored cereals and which are also free from phytotoxic and toxicological risk. The only remedy to this problem is adequate drying (see p. 237), and control of the physical environment during storage, such that re-absorption of moisture is prevented (see p. 247).

INSECT PESTS OF STORED PADDY AND MILLED RICE

Raw paddy is the least susceptible to infestation of all of the various forms of rice and in many parts of the world it is considered that paddy can be stored for long periods without deterioration. The behaviour of paddy in store will be influenced not only by grain moisture content and the physical conditions in storage, but also by such factors as maturity of the grain, the variety of paddy and harvesting and threshing methods. Husk defects, which to a large extent govern the susceptibility of paddy to insect attack may be natural (e.g. varietal or due to growing conditions), or induced (e.g. as a result of harvesting and threshing methods).

There are three primary pests of paddy, all widely distributed in the tropics, the Angoumois grain moth, the rice weevil, and the lesser grain borer; however, they are all to some degree limited in their ability to attack paddy. Other insect species, which will readily attack milled rice, are dependent upon previous attack by one of these pests, or the presence of major husk damage (such as may occur in combine harvesting or rough handling on concrete drying floors).

Milled rice is susceptible to infestation by a far greater range of insect pests than paddy because of the greater accessibility of the grain. As already indicated, the presence of broken grains in milled rice further increases susceptibility to infestation, particularly by certain moth species. Only the more important of these will be referred to here. It is often claimed that parboiled milled rice is less susceptible to infestation than raw milled rice; it has been suggested that it is probably due to the toughness imparted to the grain by the parboiling process.

Sitotroga cerealella (Lepidoptera: Gelechiidae)
Angoumois grain moth (Fig. 101)

This species commonly infests paddy in the field before harvest. It is a small (5—7 mm in length, at rest), pale yellowish-brown moth with one or two small black dots on the forewing. The apex of the hindwing, which has a very obvious fringe of hairs, is sharply pointed.

The eggs are laid, singly or in clumps, on the surface of grains and there seems little doubt that the first stage larva is able to bore through the intact husk of many varieties of paddy. The entire larval and pupal periods are passed within the grain, so that the only stage normally seen is the adult moth. Before pupation the larva prepares the way, leaving only a thin layer of seed coat intact, for the

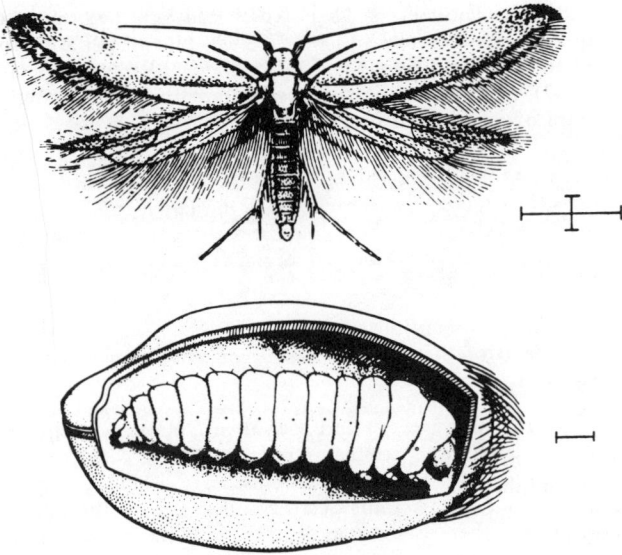

Fig. 101. The Angoumois grain moth *Sitotroga cerealella*, above, adult; below, mature larva within grain. (ICI)

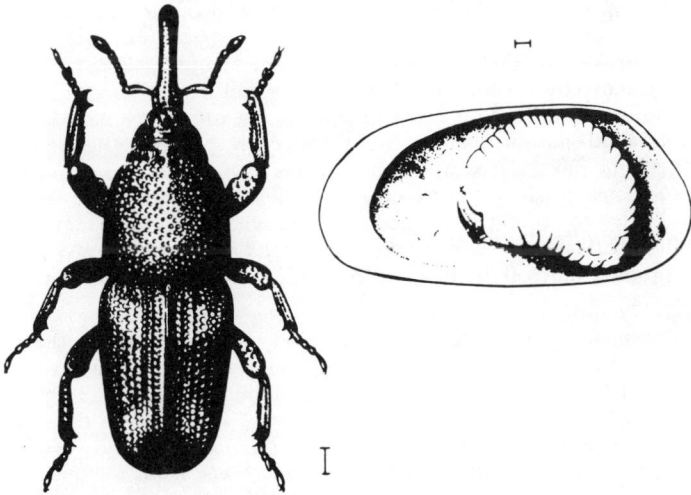

Fig. 102. The rice weevil *Sitophilus oryzae*, adult and larva inside grain. (ICI)

moth to push its way out. The life cycle is completed in five weeks at 30°C. Despite the ability of S. *cerealella* to penetrate the husk its potential as a storage pest seems to be very much influenced by the method of storage. It may be regarded primarily as a pest of bulk-stored paddy, where its activities are limited to the surface layer; of less importance in bag-stored paddy, where it is limited to the extreme periphery of a well bonded stack.

The duration of the life cycle of this species, which is a primary pest of a number of cereals, is affected by the nature of the larval food. Experimental work, much of which relates to wheat, indicates that the optimum conditions for its development are 26–30°C and 70% relative humidity.

Sitophilus spp. (Coleoptera: Curculionidae)
Rice weevil (Fig. 102)

Both *Sitophilus oryzae* (rice weevil) and *S. zeamais* (maize weevil) (formerly known as the 'small' and 'large' strains of the rice weevil, *Sitophilus (Calandra) oryzae*, respectively) are of some importance as pests of paddy and rice. *S. oryzae* is, in fact, the most serious pest of milled rice. Although both species may be found in store, there is some indication, in view of its more pronounced flight activity, that *S. zeamais* is largely responsible for pre-harvest infestation.

The two species are virtually identical and, although *S. zeamais* tends to be the larger, they can only be separated with certainty by an examination of internal characters. They are 2.5–5 mm long, dark brown to black in colour with four reddish spots on the elytra. They possess a well-defined snout and elbowed antennae characteristic of the family to which they belong.

The egg of *Sitophilus* is laid in a minute hole chewed in the grain by the female and is sealed in the hole by a waxy secretion. It has been observed that the adults of *S. oryzae* are unable to bite through the sound husk of paddy grains, even when the moisture content is high; feeding and oviposition are limited, therefore, to badly damaged grains and to kernels exposed by a separation of lemma and palea, or a split, wider than the weevil's snout (adults are often unable to emerge from grains with a husk defect which allowed oviposition, but which is too small to allow their escape). In view of the tendency for the lemma and palea to separate as a result of parboiling, it follows that parboiled paddy is more susceptible to infestation by *Sitophilus* spp. than raw paddy. However, more research is required to establish the degree to which this may be true for all varieties.

As in *Sitotroga*, the entire developmental period takes place within the grain. Both adults and larvae feed but it is the legless larva which is responsible for most of the damage. The adults live for some months, they avoid light but are active if disturbed, a characteristic which can be of advantage in the detection of infestation. Up to 150 eggs may be laid, over a period of many weeks, but with a peak of egg-laying about three weeks after emergence from the pupa.

The developmental period at the optimum conditions of 28°C and 70% relative humidity is 4–5 weeks; development is not possible below 45% relative humidity.

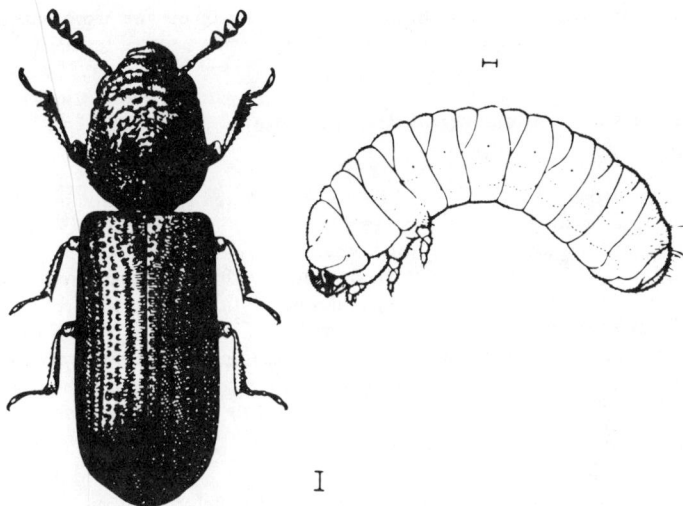

Fig. 103. The lesser grain borer *Rhyzopertha dominica*, adult and larva. (ICI)

Rhyzopertha dominica (Coleoptera: Bostrychidae)
Lesser grain borer (Fig. 103)

A brown beetle about 3 mm long, with a cylindrical body and with the head deflexed and more or less concealed from above by the prothorax, the surface of which is roughened by ridges and tubercles.

The eggs are laid loosely amongst the grains and the first instar larvae, although unable to penetrate the intact husk of rice grains, can exploit extremely narrow cracks. It has also been reported that they are able to enter the grain by boring along the centre of the rachis, but this behaviour is probably limited to immature grains. Parboiled paddy is particularly susceptible to attack by this species and it may be concluded from its lack of success on milled rice, that the presence of the husk is necessary to give the first instar larva support for penetration of the endosperm. Although the endosperm appears to be too hard for boring to commence without support, this is possible at the softer germ, as in undermilled rice.

Both adults and larvae are voracious feeders and, unlike *Sitophilus* spp., the larvae have legs and are able to feed in grain dust and to attack grains externally. The adults are long-lived and a single female may lay up to 500 eggs, over a period of three to six weeks. The optimum temperature for development is 34°C (4 weeks); the species is more tolerant of low humidity than *Sitophilus* spp., development being possible down to a minimum of 25%. Experimental work in Sierra Leone showed that *Rhyzopertha* was able to develop in undermilled rice with a moisture content of 10.3%, and that 12.4% was the optimum moisture content for this species; the minimum and optimum levels for *Sitophilus oryzae* were 11.4% and 15.0% respectively (see p. 242).

Tribolium castaneum (Coleoptera: Tenebrionidae)
Rust red flour beetle (Fig. 104)

Sometimes known as the bran bug, *Tribolium castaneum* is particularly common as a pest of rice bran; its importance as a pest of milled rice is dependent upon the degree of milling and the level of kernel breakage. It is about 3–4 mm long, rather flat, oblong and chestnut brown in colour. Up to 450 eggs may be laid at random in the produce; these hatch into slender cylindrical larvae which, like the adults, will feed on the exposed surfaces of the grains. Internal feeding is only possible following previous infestation by one of the primary pest species considered above; similarly, infestation of paddy is restricted to grains so damaged, or with major husk damage resulting from harvesting or threshing methods.

The optimum conditions for the development of *T. castaneum* are 35°C and 70% relative humidity. The duration of the life cycle varies according to the larval food, e.g. at the optimum conditions, larval development on wheat bran is from 11–16, (mean 13) days, on groundnuts 18–64 days (mean 46). Adult beetles have been known to live for as long as one and a half years.

Latheticus oryzae (Coleoptera: Tenebrionidae)
Long headed flour beetle

This beetle is 2.5–3 mm long, and is similar in appearance to *Tribolium*, but is yellow-brown and the head is a little more elongate (as its name implies). It does not differ significantly from *Tribolium castaneum* in habits and life history.

Alphitobius spp. (Coleoptera: Tenebrionidae)
Black fungus beetle

These species (*A. diaperinus, A. laevigatus*) are larger than the Tenebrionidae so far considered (5–7 mm long), somewhat oval in shape and black or dark brown in colour. They feed upon damp grain and grain residues, and their presence in rice stores is indicative of poor storage conditions involving spillage and dampness.

Oryzaephilus surinamensis (Coleoptera: Silvanidae)
Saw-toothed grain beetle (Fig. 105)

On undermilled rice (i.e. with a proportion of the oily bran remaining) this species may often be accompanied by the almost identical *O. mercator* (which shows a preference for oilseeds and their derivatives). The feeding habits of these species are similar to those of *Tribolium castaneum*, from which they may be readily distinguished both in the adult and larval stage. The adult beetle is a

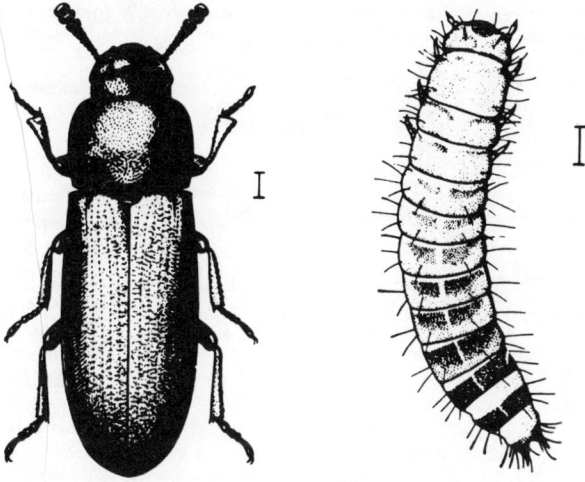

Fig. 104. The rust red flour beetle *Tribolium castaneum*, adult and larva. (ICI)

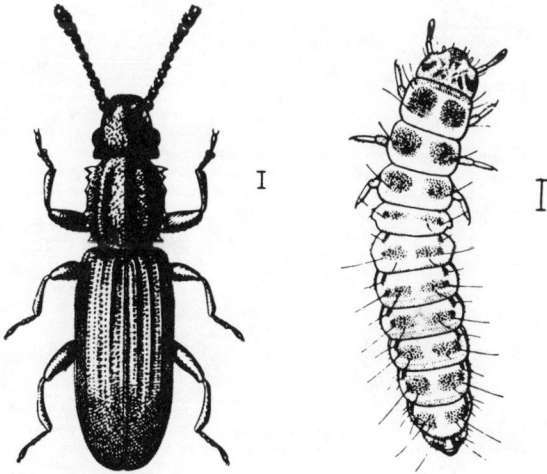

Fig. 105. The saw-toothed grain beetle *Oryzaephilus surinamensis*, adult and larva. (ICI)

little shorter and narrower than *T. castaneum;* the prothorax bears 6 large teeth along each side and has 3 ridges along the upper surface.

The optimum temperature for both species of *Oryzaephilus* is 32.5°C. Up to 300 eggs are laid over a period of about 10 weeks. Development from egg to adult under optimum conditions is completed in 4—5 weeks.

Corcyra cephalonica (Lepidoptera: Galleriidae)
Rice moth (Fig. 106)

With the wings folded along the body this moth is 12—15 mm in length; the forewings are uniformly coloured greyish-brown, without spots but with the veins slightly darkened. The head bears a projecting tuft of scales.

Particularly prevalent in milled rice with a high percentage of broken grains, infestation by *C. cephalonica* is characterised by the presence of aggregations of grains. These result from the formation, by the larvae, of silken tubes to which the grains adhere. The cocoon within which the transformation to pupa takes place may be distinguished from those of other moth species which may be found on stored rice (e.g. *Ephestra cautella;* Phycitidae) by its opaque white appearance and extreme toughness.

The adult moths are short lived and the eggs (up to 150) are laid within a few days of emergence from the pupa. The developmental period at the optimum temperature range of 28—30°C is 4—5 weeks.

Fig. 106. The rice moth *Corcyra cephalonica,* cocoon with rice grains attached, opened to show pupa inside. (Pest Infestation Control Laboratory)

STORAGE METHODS

From the foregoing remarks on insect pests of paddy and rice, it is clear that wherever long-term storage is contemplated this should be in the form of raw paddy; furthermore, in countries in which parboiling is practised this process should be so linked with rice mill throughput that storage of parboiled paddy for other than very short periods is avoided.

Storage of paddy by farmers, for food or seed, is very often in bulk (Fig. 107), in locally constructed containers which may be made from a range of materials. These are usually raised above ground in some way, primarily to prevent damage by rodents; deterioration resulting from upward movement of moisture through the soil is also prevented. The advantages of bulk storage in properly constructed containers, from the pest control point of view, will be seen from the next section, and this method can be recommended for paddy which has been adequately dried. In the use of metal bins, problems of condensation may occur, even with properly dried paddy, unless some attempt is made to shade the metal surface from direct sunlight. This is particularly so in climatic areas where there is a high day temperature followed by a sharp fall in temperature at night.

The use of a system of bulk storage, vertical or horizontal, which incorporates facilities for aeration of the paddy will to some extent alleviate problems of moisture movement through equalising grain temperature throughout the bulk; this will also control localised heating such as may occur through the existence of pockets of infested or high moisture grain. For certain varieties of rice it is considered by some authorities that it is necessary to aerate in order to maintain quality.

In the rice trade it is generally considered that ventilation is necessary if milled rice is to be stored without loss in quality, and storage, after removal of the husk, is almost invariably in bags. In considering the various ways in which rice stored in bags can be protected against attack by insect pests the importance of good warehouse hygiene cannot be over emphasised. Any portion of a store which is to receive a consignment of rice should be thoroughly swept and the sweepings burned. The walls should be brushed down and, whenever possible, sprayed with insecticide immediately prior to stacking and subsequently at regular intervals.

In order that good warehouse hygiene may be maintained and effective pest control carried out, warehouses should, preferably, be of concrete or cement block construction; rendered inside and out to provide a smooth interior finish without cavities or projections. There should be a ceiling contiguous with the walls, constructed in such a way as to ensure gas-tightness, and ventilation between this ceiling and the roof. Ventilation of the storage space should be kept to a minimum and should be such that complete ventilation control is possible, allowing it to take place only during periods of low ambient relative humidity or as required for working conditions. Ventilation openings should be fitted with fine mesh screening capable of excluding flying insects. The floor should be constructed 1.2 m above ground level and should incorporate a water-vapour barrier of bitumen (poured hot to give a continuous layer not less than

Fig. 107. Farmer's storage container for paddy in bulk (Bangladesh). (Pest Infestation Control Laboratory)

Fig. 108. Wooden pallet dunnage. (Pest Infestation Control Laboratory)

1 cm thick), bituminous roofing felt or 500 gauge polythene sheet; externally there should be adequate concrete drainage with concrete splash aprons. If these latter requirements are not fulfilled, bags should be stacked upon wooden pallet dunnage (Fig. 108) in order to prevent the uptake of moisture resulting from the upward movement of water vapour through the floor from the soil beneath.

However dry a floor may look, unless it has been treated in the manner described, this upward movement of water vapour will occur.

If wooden dunnage is used (in order that air movement beneath the bags may disperse the water vapour) it is most important that it should be lifted as soon as the bags have been removed from it so that any spillage, which would otherwise harbour dangerous insect pests, may be removed from the floor, and the dunnage thoroughly cleaned and, if necessary, treated with insecticide.

Bags should be built into stacks of uniform size (Fig. 109) in order to facilitate fumigation under gas-proof sheets. It should be remembered that it is easier, and cheaper, to treat a small stack should it become infested, than to treat a large stack should part of it become infested. In order to allow regular inspection and the application of pest control measures when necessary, stacks should be built at least 1m away from walls and with gangways of similar dimension. Stacking should be in such a way that the principle of 'first in, first out', can be followed.

Rice which is free from infestation should not be stacked close to existing stacks if these are infested, otherwise cross-infestation will occur. If shortage of space necessitates close stacking the outside surfaces of existing stacks should be

Fig. 109. Stacking bagged rice. (Note size of stack and adequate gangway between adjacent stacks). (Pest Infestation Control Laboratory)

sprayed before the new rice is received. The problem of cross-infestation of non-infested stocks, or re-infestation following fumigation can be overcome through the use of a mechanical barrier such as plastic sheeting. This technique, sometimes referred to as 'permanent plastic sheeting' has proved successful in experimental work carried out in a number of countries. A stack of produce is fumigated under gas-proof sheets on receipt into store and the sheet is left in position. Protection from crawling insects at the point of contact of the sheet with the floor, is provided by bedding the edge of the sheet in a suitable insecticidal dust. It must be emphasised that unless paddy or rice is adequately dry, the use of this technique is liable to lead to problems of moisture condensation. Fumigated produce may also be protected from reinfestation by draping cotton sheets, sprayed with insecticide, over the stack.

Insect control methods

Because of hazards to health, both of man and domestic animals, the use of only a very limited range of insecticides is permitted in stored products insect control. As indicated in Table 20 the number of these which may be applied directly to foodstuffs is limited. Tolerance limits for the use of insecticides and fumigants exist in most countries and it is most important that these should be adhered to.

Many of the major insect pests of stored rice are now frequently found to be resistant to the commonly used insecticides. For example, resistance to lindane

TABLE 20. SUMMARY OF INSECTICIDES SUITABLE FOR USE AGAINST STORED PRODUCT INSECT PESTS

Insecticide	Type of treatment			
	Space	Structure	Stack	Admixture with produce
Pyrethroid				
Pyrethrum and synthetics	X		X	X
Organochlorine				
Lindane		X	X	X
Organophosphorus				
Malathion	X	X	X	X
Dichlorvos	X			
Pirimiphos-methyl	X	X	X	X
Fenitrothion		X	X	
Bromophos		X	X	
Tetrachlorvinphos		X	X	

Note: Premium or refined grades of insecticides should always be used and the manufacturer's directions for dilution and application should be followed. Certain countries have legislation which restricts the type and amount of insecticide that may be mixed with, or used on commodities.

and malathion has been recorded in strains of *Tribolium castaneum, Sitophilus oryzae, Rhyzopertha dominica* and *Oryzaephilus surinamensis*. Resistance problems arise when the insecticide treatments which have been applied fail to kill all the insects. The survivors of the treatment, which possess a greater than average tolerance to the insecticide, if allowed to breed, may rapidly develop into a resistant population. There is too the risk that resistant infestations may be introduced into the store (on produce, equipment or dirty sacks). These insects may also survive the application of insecticidal treatments. Because of resistance it is now even more essential both to inspect all produce and equipment entering storage and to ensure that the treatments applied give complete control of infestations. Post-treatment inspections of commodities and stores are recommended.

Treatment of structure of buildings

The floor, walls and roof structure of storage premises should, after brushing down, be sprayed with insecticide. A pressurised knapsack or motorised knapsack sprayer is suitable for this purpose. Treatment should be applied to empty stores prior to stacking of produce and subsequently at regular intervals. The frequency of retreatments will depend upon the infestation present, the chemical used, the climatic conditions and the surface to which it is applied. Residual sprays of wettable powders or emulsions of malathion, pirimiphos-methyl, lindane, fenitrothion, bromophos or tetrachlorvinphos are suitable. Manufacturer's recommendations for dilution and application should be followed. Usually sprays are applied at not more than 2% a.i. and at a rate of about 5 l/100m^2. When spraying an absorbent surface such as a brick or cement-faced wall, a wettable powder is preferable to an emulsion because the solid particles are filtered out on the surface, whilst the water is absorbed into the wall.

Treatment of paddy and rice

If paddy or rice is taken into store in an infested condition the only effective method of disinfestation is by means of fumigation. In the case of produce in bags, this will have to be carried out under gas-proof sheets, unless the store is gas-tight or a special fumigation chamber is available. The sheets, should be of such a size that one is sufficient to cover a stack (hence the need for a uniform size of stack, see p. 249), but where this is not possible two or more sheets may be joined by allowing a three-foot overlap which is rolled and clipped. Fumigation does not impart any lasting protection to the produce and re-infestation will quickly occur unless preventive measures are taken. Reference has already been made to the possible use of a mechanical barrier; alternatively, after fumigation of the produce stack surfaces may be sprayed using an appropriate insecticide.

It is sometimes advantageous to use a space spray at regular intervals to prevent reinfestation of fumigated stores. Aerosol sprays or insecticidal fogs of pyrethrum or dichlorvos are particularly effective against moths. However space sprays are only suitable for use in relatively well-sealed stores and where escape of insecticide can be prevented. An automatic mist sprayer using dichlorvos can be controlled by a time clock.

An advantage of storage in bulk in silos (e.g. with paddy) is that, provided the structure is sufficiently gas-tight, fumigation is possible and re-infestation should not occur. However, in the case of small to medium sized bins, entry of insects via the roof (if such exists) is very often possible; under such circumstances protection may be achieved through the admixture of lindane or malathion dust with the surface grain (i.e. paddy). Rodent control is facilitated by storage in silos and in warehouses specially constructed for storage in a horizontal bulk, but in the latter case insect control operations are difficult.

Certain insecticidal dusts may be directly admixed with paddy to give protection from insect attack. For example malathion may be admixed at the rate of 10 ppm (10 mg a.i./1 kg of paddy or 78 g of 1% dust per 78 kg bag of paddy). This treatment cannot, however, be recommended for use with milled rice because of the danger of excessive contamination by insecticide residues. Before consumption, paddy undergoes either parboiling, drying and milling, or just milling, so that there is little chance of undue contamination of the milled product.

Recommended fumigants

1. Methyl bromide: This is the most widely used fumigant, but it is also highly toxic to man and should therefore be applied only by personnel specially trained in its use. Its penetration properties make it suitable for the treatment under gas-proof sheets of large stacks of bagged rice; the recommended dosage is 24 g methyl bromide/m^3 for an exposure period of 24 hours.

2. Phosphine: In the presence of water vapour, tablets or pellets of aluminium phosphide yield the gas phosphine. An advantage of this method is the ease of application of the fumigant. Tablets may be added (at a calculated rate) to the stream of paddy entering a bin; or they may be distributed around the base and on the top of a stack of bagged paddy or rice, which is then covered with gas-proof sheeting. The normal rate of application is 5 tablets (or 25 pellets) per ton, the fumigation period being not less than five days.

3. Chlorinated hydrocarbons: The most commonly used of these 'liquid fumigants' as they are often called, is a mixture of 3 parts ethylene dichloride to 1 part carbon tetrachloride (by volume). Because it is applied as a liquid it is most appropriate for the treatment of paddy in small bins, application to the surface grain being carried out through the use of a stirrup pump or watering can. Small stacks of paddy or rice can also be quite easily fumigated, under gas-proof sheets, using this mixture. Although care must be taken in handling any volatile chemical, the operator hazards are very much less in the case of liquid fumigants than with methyl bromide. The normal dosage rate for the 3:1 mixture referred to above is 2.8 l/t of produce for a fumigation period of 48 hours.

Treatment of empty bags

An important factor in the spread of infestation which is all too often ignored is the re-use of second-hand bags without any attempt at disinfestation. A simple technique for the treatment of small quantites is fumigation using the mixture of ethylene dichloride and carbon tetrachloride considered above. Twenty bags can be rolled together and placed vertically in a large dustbin or 200 *l* drum with a tightly fitting lid and fumigant, at the rate of 0.5 *l*/45 kg of sacks for 70 hours, applied evenly to the top surface using a watering can. Aluminium phosphide tablets may also be used. Disinfested and new bags should always be kept in a special store situated away from produce warehouses in order to minimise the risk of infestation before use.

Rodent control methods

Damage by rodents can be a serious problem, particularly at the farmer and trader levels of storage. Direct damage to grain through the feeding activity of rats and mice is important, but a much larger quantity of grain is contaminated with their faeces and urine. Damage to containers such as jute bags can also be serious.

The species of rodents found in food stores are generally the three world-wide commensals: the Norway rat, the ship rat and the house mouse. Occasionally stores on farms may be entered by field rats of various species according to the region in which the store is situated. But field rats are not well adapted to living in buildings and rarely cause serious damage.

These are two approaches to the prevention of damage by rodents: either regular disinfestation in and around the store, or through rodent-proofing of the store to prevent the entry of these pests. Taking the long-term view proofing is, of course, the more satisfactory approach. But proofing is not always practicable and regular poisoning may be necessary to keep rodents in check (Figs. 110—112).

Fig. 110. Norway rat *Rattus norvegicus,* a pest of stored rice. (Pest Infestation Control Laboratory)

Fig. 111. House mouse *Mus musculus,* a typical pest of grain stores. (Pest Infestation Control Laboratory)

Fig. 112. Rats killed after warehouse fumigation. (Note protective mask worn by operator). (Pest Infestation Control Laboratory)

(see Rodents, p. 219). Chronic or acute poisons can be used in stores but great care must be taken to prevent contamination of foodstuffs. Poisoned water may be useful where the normal cereal baits fail to attract rodents. It should be remembered that any fumigant which kills insects will also kill rodents.

Bibliography

ASHMAN, F., WEBLEY, D. J. and ACHILLIDES, N. S. (1974). Space treatment of warehouses using dichlorvos. *European Plant Protection Organization Bulletin* 4 (4): 429—445.

BALDACCI, E. and CORBETTA, G. (1964). Ricerche sulla microflora della cariossidi di riso dopo conservazione in magazzino e in condizioni sperimentali. (Research on the microflora of rice kernels after storage in warehouses and in experimental facilities). *Il Riso* 13 (1): 79—88. (English summary).

BREESE, M. H. (1960). The infestibility of stored paddy by *Sitophilus sasakii* (Tak.) and *Rhyzopertha dominica* (F.). *Bulletin of Entomological Research* 51 (3): 599—630.

BREESE, M. H. (1963). Studies on the oviposition of *Rhyzopertha dominica* (F.) in rice and paddy. *Bulletin of Entomological Research* 53 (4): 621—637.

CHRISTENSEN, C. M. and LOPEZ, L. C. (1963). Pathology of stored seeds. *Proceedings of the International Seed Testing Association* 28: 701—711.

CRAUFURD, R. Q. (1963). Research into the improvement of milling and drying techniques. *L'Agronomie Tropicale* 18 (8): 844—846.

DYTE, C. E. (1974). Problems arising from insecticide resistance in storage pests. *European Plant Protection Organization Bulletin* 4 (3): 275—289.

FERNANDO, H. E. (1959). Storage loss of paddy due to *Sitotroga cerealella* and its control. *International Rice Commission Newsletter* 8 (1): 20—25.

GHOSE, R. L. M., GHATGE, M. B. and SUBRAHMANYAN, V. (1960). *Rice in India*. Revised edition. New Delhi, Indian Council of Agricultural Research. 474 pp.

GLINSUKON, T., YUAN, S. S., WIGHTMAN, R., KITAURA, Y., BUCHI, G., SHANK, R. C., WOGAN, G. N. and CHRISTENSEN, C. M. (1974). Isolation and purification of Cytochalasin E and two Tremorgens from *Aspergillus clavatus*. *Plant Foods for Man* 1: 113—119.

HOUSTON, D. F. (Ed.) (1972). *Rice: Chemistry and Technology*. Monograph Series. Vol. 4. American Association of Cereal Chemists, Inc., St Paul, Minnesota, USA. xviii + 517 pp.

HOWE, R. W. (1956). The effect of temperature and humidity on the rate of development and mortality of *Tribolium castaneum* (Herbst) (Coleoptera, Tenebrionidae). *Annals of Applied Biology* 44 (2): 356—368.

LUCAS, F. V., MONROE, P., NGA, P. V., TOWNSEND, J. F. and LUCAS, F. V. (1971). Mycotoxin contamination of South Vietnamese rice. *Journal of Tropical Medicine and Hygiene*. 74 (8): 182—184.

PREVETT, P. F. (1959). An investigation into storage problems of rice in Sierra Leone. London, HMSO. 52 pp. *Colonial Research Studies* 28.

PURCHASE, I. F. H. (Ed.) (1974). *Mycotoxins*. Amsterdam, Elsevier Scientific Publishing Co.

WRIGHT, F. N. and SOUTHGATE, B. J. (1962). The potential uses of plastics for storage with particular reference to rural Africa. *Tropical Science* 4 (2): 74—81.

PESTICIDE APPLICATION

Introduction

Various techniques have been used to apply pesticides for the control of rice pests. High volume spraying with wettable powder or emulsifiable concentrate formulations is one of the recommended methods but this necessitates mixing the chemical with large volumes of water (up to 1000 l/ha). Simple lever-operated knapsack sprayers are used so small sprayers require refilling many times (Table 21) and various adaptations (Fig. 113) have been designed to improve the speed of application and reduce contamination hazards for the operator, even though dilute sprays (0.01—0.5% a.i.) are usually applied by high volume application. The swath width can be increased by mounting a horizontal boom on the rear of a knapsack sprayer so the operator walks away from the treated crop rather than into it as with a conventional lance normally held in front of him. Reduction in contamination of the operator can also be achieved by mounting two wide angle fan nozzles on rods fixed on the back of the spray tank. Also, a vertical boom with a single impact nozzle has been developed for herbicide application (Fig. 114). With applications of large volumes of spray, a nozzle with a large orifice is often used to spray the liquid quickly, but this results in a larger droplet size. As the nozzle is usually held above the crop, large droplets tend to be deposited on the more exposed surfaces of the plant or the spray liquid 'runs off' to the soil or irrigation water. Only a small proportion of the total spray may be retained on the foliage and, if heavy rain falls,

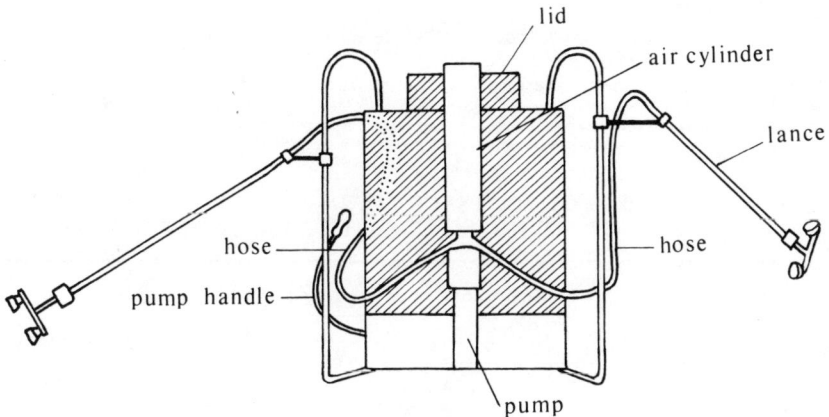

Fig. 113. Double lanced knapsack sprayer with the nozzles on fixed lances behind the operator. (L.H.)

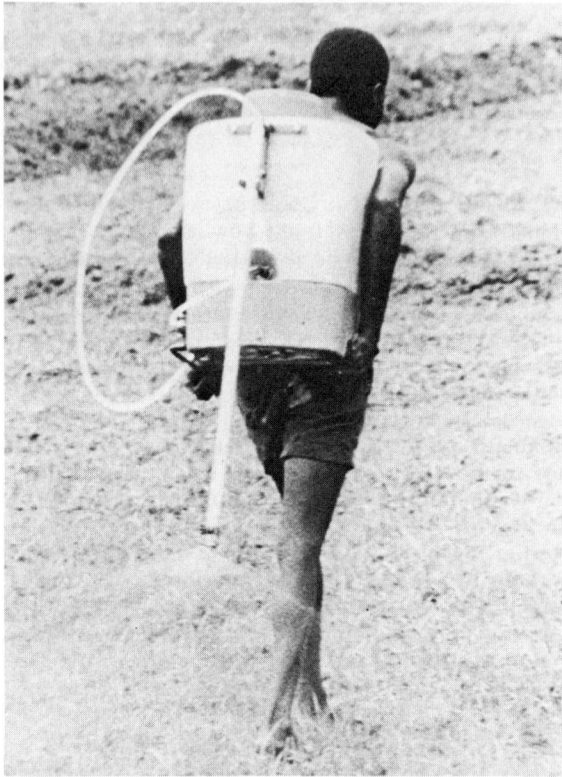

Fig. 114. Vertical boom with single impact nozzle for herbicide application attached to Cooper Pegler CP3 knapsack sprayer. (National Institute of Agricultural Engineering)

TABLE 21. NUMBER OF KNAPSACK LOADS NEEDED FOR A RANGE OF VOLUME APPLICATION RATES

	High volume application sprayer with capacity			
l/ha	10 l		15 l	
	loads/ha	m^2/load	loads/ha	m^2/load
50	5	2000	3.3	3000
200	20	500	13.3	750
1000	100	100	66.7	150

much of the spray deposit is washed from the foliage so a repeat spray may be necessary. Also as plant growth is rapid, spraying gives little protection of the new foliage. High volume spraying is slow and laborious so has not been very popular with farmers. Even if water is readily available as on irrigation schemes, collecting the water and mixing sprays takes a considerable proportion of the total spraying time.

Low volume spraying

Application of foliar sprays is much quicker if less liquid is sprayed. 20–50 l/ha can be applied without run-off using a motorised knapsack mistblower. More recently, ultra-low volume (u.l.v.) application with less than 10 l/ha has achieved good results. Mistblowers adapted for a lower discharge rate can apply a concentrated formulation at as little as 1 l/ha for stem borer control and 2 l/ha for rice-blast disease using 8 and 6 m swaths respectively. Spraying a hectare can be completed within an hour (Table 22). Small droplets are required for good coverage so the usual gaseous energy nozzle on some mistblowers has been modified by mounting a rotary disc or propeller in the airstream (Figs. 115–117). Many mistblowers produce too wide a range of droplet sizes so the smallest droplets (<30 μm vmd) may drift from the treated area and the largest are filtered out on the nearest foliage. Initial distribution of droplets is in a narrow high velocity airstream but the energy is dissipated rapidly so final deposition is largely influenced by natural air movements. Thus when the nozzle is directed over the crop, droplets may drift over a wide swath and few droplets penetrate the crop canopy. Directing the nozzle into the crop also gives uneven coverage because the droplets are entrained in such a narrow jet of air.

TABLE 22. ULTRA LOW VOLUME APPLICATION—DISCHARGE RATE, SWATH WIDTH, WALKING SPEED AND TIME REQUIRED FOR APPLICATION

Spray rate l/ha	Swath width m	Discharge rate ml/min	Walking speed m/s	Spraying time min/ha
1	6	14.4	0.4	69.4
1	6	21.6	0.6	46.3
1	8	19.2	0.4	52.0
1	8	28.8	0.6	34.7
2	6	28.8	0.4	69.4
2	6	43.2	0.6	46.3
2	8	38.4	0.4	52.0
2	8	57.7	0.6	34.7
3	6	43.2	0.4	69.4
3	6	65.0	0.6	46.3
3	8	57.7	0.4	52.1
3	8	86.6	0.6	34.7

Fig. 115. Rotary disc atomiser type of low-volume
concentrate sprayer. (T. Takenaga, Institute of
Agricultural Machinery, Saitama, Japan, *Japan
Pesticide Information,* 1972)

The main disadvantages of motorised knapsacks are the high initial cost,
difficulties in maintenance of the two-stroke engine and excessive noise and
weight of the machine. Many farmers prefer to walk along ridges between
fields and attempt to drift spray over too wide a swath rather than walk through
paddy-fields with a heavy machine.

The main fault liable to occur with two-stroke engines is that the engine may be
difficult to start or run erratically. This problem can be reduced by regular
cleaning of the air filter and spark plug. If the float needle in the carburettor
sticks, it will require cleaning or replacement with a spare part. The petrol/oil
fuel must be carefully mixed and should be drained from the fuel tank at the end
of each day's spraying. When stopping the machine, the fuel tap should be closed
and the carburettor allowed to run dry.

More recently, light-weight centrifugal energy nozzles powered by high power
batteries have been developed for controlled droplet application (c.d.a.) spraying. Controlled droplet application is now possible as droplet size is inversely
proportional to the rotational speed of the spinning disc or cup. Large droplets
(about 300 μm vmd) are produced from a 9 cm disc rotated at 1800 rev/min.
Although designed for herbicide application to avoid drift, this sprayer can be
used with suitable formulations for applying 'liquid granules' or encapsulated

Fig. 116. Propeller atomiser type of low-volume concentrate sprayer. (T. Takenaga, Institute of Agricultural Machinery, Saitama, Japan. *Japan Pesticide Information*, 1972)

Fig. 117. Fixed edge atomiser type of low-volume concentrate sprayer. (T. Takenaga, Institute of Agricultural Machinery, Saitama, Japan. *Japan Pesticide Information*, 1972)

formulations to paddy fields. At higher disc speeds, droplets as small as 30 μm vmd can be obtained for insecticide and fungicide application. Released just above the crop, these small droplets move with the air currents within the crop canopy and are filtered out on the vertical rice stems and leaves. When initially developed, it was hoped that drift spraying over wide swaths (10 m) would be possible. However, wind velocity fluctuates considerably so narrow swaths (2—5 m) allow overlapping and greater penetration within the canopy and reduce the risk of leaving untreated areas.

The spinning disc is held 0.5—1.0 m above the crop on the leeward side of the operator as he walks across the field. At the edge of the field, he inverts the bottle to stop spraying but keeps the motor operating while he walks upwind to the start of the next swath (Fig. 118). The spray droplets are therefore carried away from the operator who walks between untreated plants. With the bottle held vertical during spraying, liquid is fed from a plastic bottle by gravity through a restrictor to the disc (Fig. 119). The restrictor is interchangeable for various flow rates, but the lower flow rates are preferred to avoid overloading the disc. At a walking speed of 1 m/s, flow rate of 0.5 ml/s and a swath of 2 m, rate of application is 2.5 l/ha. A relatively non-volatile formulation is needed at such a low volume, but up to 15 l/ha of water-based formulations have been applied with these sprayers.

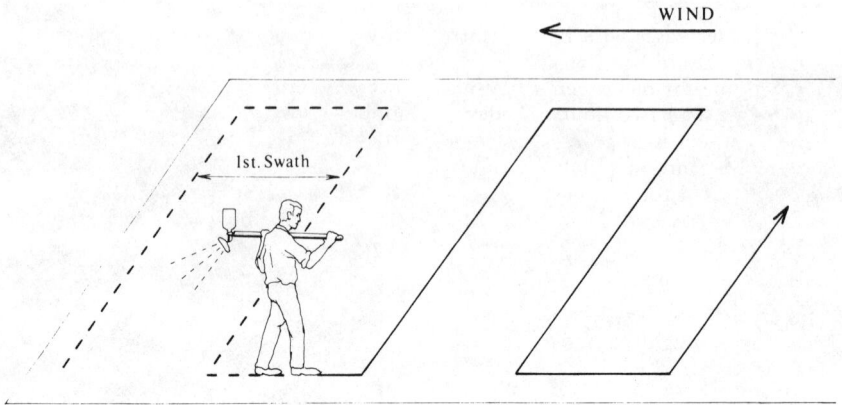

Fig. 118. Spraying with light-weight centrifugal energy nozzles powered by transistor batteries. (L.H.)

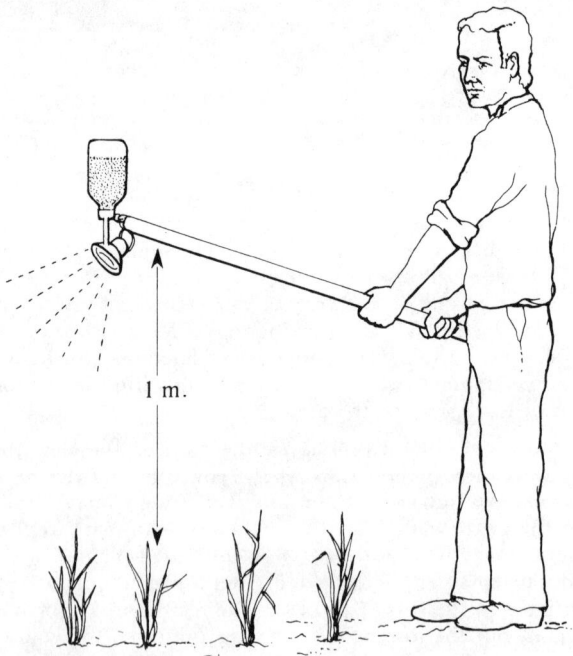

1 m.

Fig. 119. The spinning disc is held about 1 m above the crop and the pesticide is fed from a plastic bottle by gravity through a restrictor to the disc. (L.H.)

Disc speed decreases with battery fatigue so when small droplets are required, the sprayer should be operated intermittently to prolong battery life. With the 'ULVA'*, one set of sixteen 1.5 V high power batteries lasted 13 hours if used for no more than two hours per day and applying spray at the rate of 0.5 ml/s. Further research is in progress to improve these sprayers and reduce power requirements, and on suitable formulations for c.d.a. spraying. Although not extensively used for rice pest control so far, the potential for these relatively inexpensive sprayers for individual small-scale farmers is enormous.

Granule application

Over the last few decades, because of difficulties with spraying equipment, farmers have sought an easier method of applying pesticides. Granules can be broadcast by hand without mixing or the need for specialised equipment (Fig. 120).

Fig. 120. Application of granular insecticides (from Custodio, H., Sanchez, F. F., Pathak, M. D., Dyck, V. A. and Feuer, R. 1973). (L.H.)

One of the first granule formulations used was γBHC. The effectiveness of these granules against rice stem borers is partly due to a fumigant effect but some of the pesticide is also absorbed from the irrigation water (systemic action). Movement of pesticide is also by capillary action as leaves are closely packed together at the base of the plant. Application with granules tends to be more effective than sprays under wet conditions, but comparatively little of the pesticide reaches the upper leaves so granules of many insecticides are not very effective against plant and leafhoppers.

Retention of granules on the foliage can be improved by using micro- or fine granules (Table 23), which do not drift like the very small particles in dust formulations. Even with micro-granules, the proportion retained on the plants is still low.

---

*Manufactured by Micron Sprayers Ltd., Bromyard, Herefordshire.

TABLE 23. DEPOSITS ON PLANTS WHEN THREE FORMULATIONS OF INSECTICIDES WERE APPLIED WITH A BOOM TYPE BLOW HEAD ATTACHED TO A MOTORISED KNAPSACK DUSTER

Formulation	Dosage kg/ha	Deposits on plants mg/hill	Insecticide on plants (%)
Fine granule	35	38.6	24.3
Granule	44	7.9	4.1
Dust	46	31.0	15.0

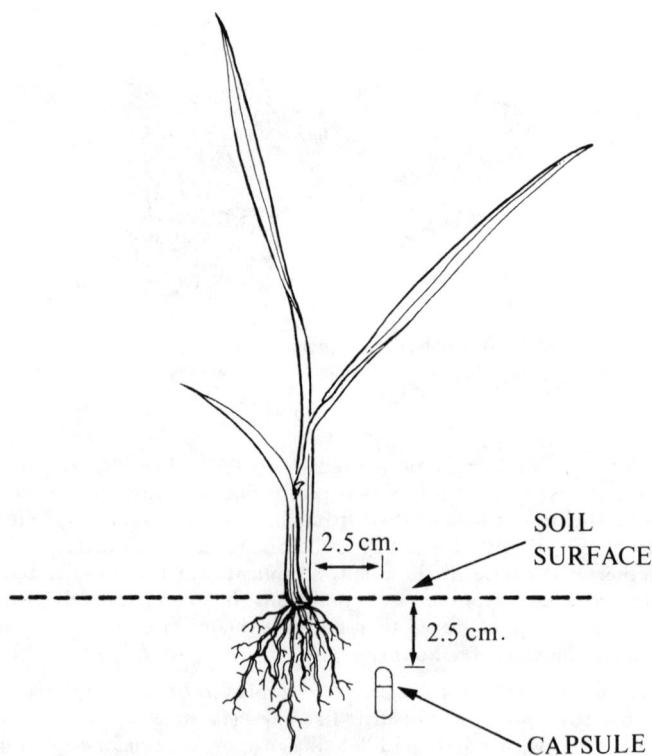

Fig. 121. The placement of gelatine capsules for insecticide treatment (from Encarnacion, D. and Dupo, H., 1974). (L.H.)

A major disadvantage of granules is that only a small proportion of the active ingredient reaches the correct site of action. Much of the pesticide is lost in the irrigation water, so granule application requires more active ingredient per unit area. The usual rate of many insecticides is 2 kg/ha active ingredient in most granule formulations, a large amount of inert material is also applied. This increases both the basic cost of the formulated material and also transport costs which are greater than for the concentrated formulations used in sprays. In addition, the farmer must carefully control his irrigation water to prevent loss of pesticide in the outflow of water. Care must also be taken to avoid killing fish which are highly susceptible to certain pesticides.

Root zone application

Placing granules in the irrigation water was avoided by their placement in the bottom of furrows of direct-seeded rice or soaking the roots of seedlings in a suspension of wettable powder immediately prior to transplanting. The addition of a sticker (2% methyl cellulose or 4–6% corn, cassava or wheatflour) improved adherence of insecticide to the roots. This root coat application gave good control of soil-inhabiting insects but protection is limited to the early part of the season and supplementary treatments were necessary.

Improved persistence by protecting the insecticide from leaching and other losses due to the flow of paddy water as well as reducing the effects of volatilisation, heat and sunshine, was achieved by placement of the insecticide inside cut paper straws punched with holes and placed 2.5 cm below the soil surface and 2.5 cm away from the hill, three to five days after transplanting. Instead of paper straws, gelatine capsules (Fig. 121) have been tried, and to reduce costs, insecticide granules have been mixed in clay and made into pills which were hardened in the sun or applied directly. This technique appears to be less hazardous to parasites and predators, and by keeping the chemical in the root zone, absorption by the plant is greater. Results achieved with chlordimeform applied by different techniques are shown in Fig. 122.

Fertilizer is also applied in the root zone so the inert carrier for the insecticide may not be necessary if the chemical is compatible with the fertilizer and can be mixed with it without causing phytotoxic damage to the roots. Instead of granules, both fertilizer and insecticide can be formulated in a paste and injected to the correct depth. These possibilities are being investigated.

The application techniques described above have various disadvantages. A major problem is the control of the brown planthopper, *Nilaparvata lugens*. This pest is difficult to control because both adults and nymphs live mostly at the base of plants where it is shaded and moist (Fig. 123). Populations can increase rapidly especially on high-tillering varieties and in areas with irrigation throughout the year. Serious infestations occur when plants reach the heading stage, then foliage of adjacent hills covers the area to such an extent that penetration of spray droplets is limited. With high volume spraying, the lance should be held so that the nozzle is directed sideways between the plants (Fig. 124) but this makes spraying even more laborious. Ideally, control is needed simultaneously over a wide area to reduce possible re-infestation. Only certain of the most toxic

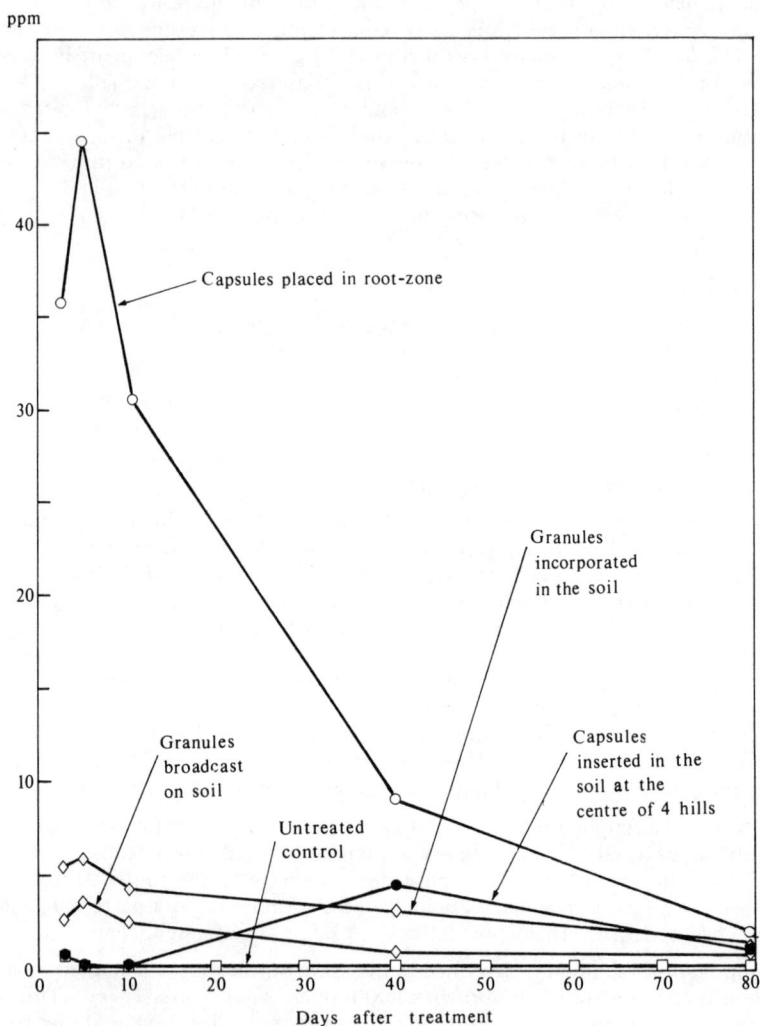

Fig. 122. Concentration of chlordimeform and its metabolites in rice shoots as influenced by different methods of application (from Encarnacion, D. and Dupo, H., 1974). (L.H.)

Fig. 123. Position of brown planthoppers on rice at heading stage (from Custodio, H. *et al.*, 1973). (L.H.)

Fig. 124. Left, correct method, spray for brown planthoppers by horizontal direction of the nozzle between rows directed towards the lower parts of the plants where the brown planthoppers live. Right, wrong method, do not spray the tops of the plants as you would for green leafhopper control because the brown planthopper lives near the base of the plants (Custodio, H. *et al.*, 1973). (L.H.)

insecticides such as carbofuran have been effective using granules and root zone application against brown planthopper. The mammalian toxicity of such materials poses problems of distribution and packaging in low technology areas.

Other application techniques

Effective control can be obtained by mixing insecticide in a high spreading oil (h.s.o.) applied at regular intervals. The oil spreads over the surface of the paddy water. As the h.s.o. formulation contains a high concentration of active ingredient, it compares favourably with the bulky low concentration of granules.

Rice can be sprayed by aircraft especially with u.l.v. applications of relatively non-volatile formulations. 300,000 ha were sprayed in Indonesia during 1968/69 and other large areas have been similarly treated. Aerial application is particularly suited to the control of major outbreaks quickly, provided suitable aircraft are available for immediate use. Use of aircraft for routine pest control is more expensive than ground techniques and requires a high standard of cultivation over the whole area for maximum economic benefit. Where sowing dates and fertilizer practice vary between fields, intensity of pest attack is also likely to vary so overall application is not economic.

Chlordimeform e.c. has been applied as a spot treatment by means of a 'Paddy Spotter' instead of granule application. This is a simple volumetric gadget, fitted to a small pesticide container (1 l capacity) which is tilted backwards and then slightly forwards to apply 1.5 ml of formulation once every 2.5 m across the field, thus with 4 m swaths, 1.5 l/ha is applied.

Safety precautions

Protective clothing should be worn by those applying pesticides as recommended by the appropriate authorities, and according to the toxicity of the pesticide and local conditions. Toxicity hazards are reduced by granule application provided the granules do not disintegrate to a fine powder; fine dusts can be inhaled as well as drift from treated areas. Unfortunately, many of the granule formulations used in rice contain the most toxic insecticides with a mammalian toxicity of less than 50 mg/kg. When applying these, operators should wear boots and gloves.

When ordinary knapsack sprayers with a lance are used, the lance should be held on the leeward side rather than in front of the operator.

Operators of motorised mistblowers as well as other sprayers should be warned that they can be contaminated with spray when a change of wind direction occurs or they attempt to spray upwind. Despite the strong airstream at the nozzle, droplets may blow back on the operator.

When concentrated sprays are applied at ultra-low volume, dermal contamination should be reduced as far as possible by wearing overalls. When proper protective clothing is not available, the least toxic chemicals may be applied but an old pair of long trousers and long-sleeved shirt should be worn and washed immediately after spraying.

The future

The various techniques for pesticide application described have a number of limitations. Studies on the behaviour of rice pests may indicate more precisely where the pesticide is needed. Development of improved spray techniques with greater control of droplet size, together with improvements in formulation such as encapsulation to reduce losses of pesticide due to volatilization and rain-washing, may result in greater efficiency in the deposition on the most appropriate target. Hopefully, such a development may allow pesticides to be used more selectively in an integrated control programme and reduce undesirable effects on non-target organisms such as fish.

Bibliography

BALS, E. J. (1975). Development of a c.d.a. herbicide handsprayer. *PANS* 21 (3): 345–349.

CADOU, I. (1959). Une rampe portative individuelle pour la pulverisation a faible volume. *Coton et Fibres Tropicales* 14: 47–50.

CLAYPHON, J. E. and MATTHEWS, G. A. (1973). Care and maintenance of spraying equipment in the tropics. *PANS* 19 (1): 13–23.

CLAYPHON, J. E. and THORNHILL, E. W. (1974). Spinning disc nozzle adaptation for mistblowers. *British Crop Protection Council Monograph* 11: 281.

CUSTODIO, H., SANCHEZ, F. F., PATHALE, M. D., DYCK, V. A. and FRIER, R. (1973). *Brown planthopper control.* Leaflet issued by Interagency Committee for Pest Control in Lowland Rice, Philippines.

ENCARNACION, D. and DUPO, H. (1974). Recent advances in insecticidal control of insect pests of rice. *International Rice Research Conference* IRRI, Philippines.

FERNANDO, H. E. (1956). A new design of sprayer for reducing insecticide hazards in treating rice crop. *FAO Plant Protection Bulletin* 4: 117–120.

HATAI, N. (1972). Performance guide for ULV ground spraying in paddy field. *Japan Pesticide Information* 11: 35–36.

IWATA, T. (1973). Rice insect control by fine granular formulation of insecticides in Japan. *Japan Pesticide Information* 14: 23–26.

LIM, G. S. (1973). Control of rice insects using ULV concentrate and high spreading oil insecticides in Malaysia. *Malaysian Agricultural Journal* 49: 122–130.

MATTHEWS, G. A. and CLAYPHON, J. E. (1973). Safety precautions for pesticide application in the Tropics. *PANS* 19 (1): 1–12.

MATTHEWS, G. A. and MOWLAM, M. (1974). Some aspects of the biology of cotton insects and their control with ULV spraying in Malawi. *British Crop Protection Council Monograph* 11: 44–52.

SAMA, S. and VAN HATTEREN, P. (1974). Insecticide applied in mudballs to lowland rice. *The Rice Entomology Newsletter* 1: 5–6.

TAKENAGA, T. (1972). Knapsack Type LV Concentrate (ULV) Sprayer. *Japan Pesticide Information* 13: 5–10.

APPENDIX

CHECK LIST OF DISEASES

Species	Synonymy	Common name
Cercospora oryzae Miyake		Narrow brown leaf spot
Cochliobolus miyabeanus (Ito & Kuribay.) Drechsler ex. Dastur	*Drechslera oryzae* (Breda de Haan) Subram. & Jain (imperfect state) *Helminthosporium oryzae* van Breda de Haan	Brown spot Sesame spot Helminthosporiosis Seedling and leaf blight
Corticium rolfsii Curzi	*Sclerotium rolfsii* Sacc. (imperfect state) *Hypochnus centrifugus* Tul. *Corticium centrifugum* (Lév.) Brés. *Botryobasidium rolfsii* Venkat. *Pellicularia rolfsii* (Sacc.) West	Seedling blight
Corticium sasakii (Shirai) Matsumoto	*Rhizoctonia solani* Kuhn (imperfect state) *Corticium vagum* Berk & Curt. and Park & Bertus *Nypochnus sasaki* Shirai *Pellicularia filamentosa* (Pat.) Rogers *Thanatephorus cucumeris* (Frank) Donk	Sheath blight
Entyloma oryzae H. & P. Sydow	*Ectostroma oryzae* Sawada *Sclerotium phyllachoroides* Hara	Leaf smut
Gibberella fujikuroi (Saw.) Ito	*Fusarium moniliforme* Sheld (imperfect state)	Bakanae disease and foot rot
Helminthosporium sigmoideum Cav. var. *irregulare* Cralley & Tullis	*Nakataea irregulare* Hara *Vakrabeeja sigmoidea* (Cav.) Subram. var. *irregulare* (Cralley and Tullis) Shoemaker	Irregular stem rot
Magnaporthe salvinii (Catt.) Krause & Webster	*Nakataea sigmoidea* (Cav.) Hara (imperfect state) *Sclerotium oryzae* Cav. (sclerotial state) *Vakrabeeja sigmoidea* (Cav.) Subram.	Stem rot

Species	Synonymy	Common name
Pyricularia oryzae Cav.	*Piricularia oryzae* Cav.	Blast
Rhizoctonia oryzae Ryker & Gooch		Bordered sheath spot
Sphaerulina oryzina Hara	*Cercospora oryzae* Miyake (imperfect state)	Narrow brown leaf spot
Tilletia barclayana (Bref.) Sacc. & Syd.	*Neovossia barclayana* Bref. *Tilletia horrida* Takahashi *Neovossia horrida* (Takahashi) Padwick & Azamatullah Khan	Smut Black smut
Ustilaginoidea virens (Cke.) Tak.	*Ustilago virens* (Cke.) *Tilletia oryzae* (Pat.) *Ustilaginoidea oryzae* (Pat.) Brefeld *Sphacelotheca virens* Omori *Claviceps virens* (Cke.) Sakurai ex Nakata *Claviceps oryzae-sativae* Hashioka	False smut
Xanthomonas oryzae (Uyeda & Ishiyama) Dowson	*Pseudomonas oryzae* Uyeda & Ishiyama *Xanthomonas oryzae* Uyeda & Ishiyama *Bacterium oryzae* Uyeda & Ishiyama Nakata	Bacterial leaf blight
Xanthomonas translucens (Jones, Johnson & Reddy) Dowson f. sp. *oryzicola* (Fang *et al.*) Bradbury	*Xanthomonas oryzicola* Fang *et al.* *Xanthomonas translucens* f. sp. *oryzae* Pordesimo	Bacterial leaf streak

CHECK LIST OF INSECTS

Pest	Some synonyms or misidentifications	Common name
Agromyza oryzae (Munukata)	*Oscinis oryzella* (Mats.) *Agromyza oryzae* Hend.	Rice leaf miner
Alphitobius diaperinus (Panz.)		Black fungus beetle
Alphitobius laevigatus (F.)	*Alphitobius piceus (Ol.)*	Black fungus beetle
Atherigona exigua Stien		Rice seedling fly, bibit fly, stem-mining maggot
Baliothrips biformis (Bagn.)	*Chloethrips oryzae* (Williams) *Thrips oryzae* Williams	Rice thrips, paddy thrips
Cadra cautella (Wlk.)		Tropical warehouse moth
Chaetocnema concinnipennis Baly		
Chaetocnema pulla Chap.		
Chilo agamemnon Blesz.		Small purple-lined borer
Chilo auricilius Dudgn.	*Chilotraea auricilia* (Dudgn.) *Diatraea auricilia* Dudgn.	Gold-fringed rice borer
Chilo partellus (Swinh.)	*Chilo zonellus* (Swinh.)	Maize or sorghum stem borer
Chilo plejadellus Zk.		American rice stem borer
Chilo polychrysus (Meyr.)	*Diatraea polychrysa* Meyr. *Proceras polychrysa* (Meyr.) *Chilotraea polychrysa* (Meyr.)	Dark-headed striped borer
Chilo suppressalis (Wlk.)	*Crambus suppressalis* Wlk. *Chilo simplex* (Butl.) *Chilo oryzae* Fletcher	Asiatic rice borer, pale-headed striped rice borer
Chironomus cavazzai Kieff		
Chironomus tepperi Skuse		
Chironomus thummi (Kieff.)		
Chlorops oryzae (Mats.)		Rice stem maggot, rice shoot fly
Cletus trigonus Thnb.		Slender rice bug, paddy ears
Cnaphalocrocis medinalis (Gn.)	*Cnaphalocrocis jolinalis* Led.	Rice leaf folder, rice leaf roller, grass leaf roller
Colaspis flavida (Say)		Grape colaspis
Corcyra cephalonica (Stnt.)		Rice moth
Cricotopus sylvestris (F.)		
Cricotopus trifasciatus (Panz.)		
Dicladispa armigera (Ol.)	*Hispa armigera* Ol.	Rice hispa, paddy hispa, rice leaf beetle

273

Pest	Some synonyms or misidentifications	Common name
Dicladispa gestroi (Chap.)		
Dicladispa viridicyanea (Kraatz)	*Chrysispa viridicyanea* Kraatz	
Diopsis thoracica Westw.		Stalk-eyed borer
Diploxys fallax Stål		Rice shield bug
Dysmicoccus boninsis (Kuw.)		
Dysmicoccus brevipes (Ckl.)		Pineapple mealybug
Echinocnemus oryzae Mshl.		Paddy root weevil
Eusarcoris inconspicuus (H.-S.)		Chinche del arroz
Geococcus oryzae (Kuw.)	*Ripersia oryzae* Kuw.	
Gryllotalpa africana (P.de B)		Mole cricket
Haplothrips aculeatus (F.)		
Heterococcus rehi (Ldgr)	*Ripersia sacchari* Green *Ripersia oryzae* Green *Ripersia rehi* Ldgr *Tychea rehi* (Ldgr)	Rice mealybug
Heteronychus oryzae Britten		
Hieroglyphus banian (F.)		Large rice grasshopper
Hieroglyphus nigrorepletus (I. Bol.)		Large rice grasshopper
Hydrellia griseola (Fall.)		Rice leaf miner, smaller rice leaf miner
Hydrellia philippina Ferino		
Hydrellia sasakii Yuasa & Isitani		Paddy stem maggot
Laodelphax striatellus (Fall.)	*Delphacodes striatella* Fall.	Small brown planthopper
Latheticus oryzae Waterhouse		Long-headed flour beetle
Latoia bicolor (Wlk.)	*Parasa bicolor* Wlk.	Slug caterpillar, nettle grub
Leptocorisa acuta (Thnb.)	*Leptocorisa varicornis* (F.)	Rice bug
Leptocorisa chinensis Dallas	*Leptocorisa corbetti* China	Rice bug
Leptocorisa costalis (H.-S.)		Rice bug
Leptocorisa discoidalis Wlk.		Rice bug
Leptocorisa lepida Bredd.		Rice bug
Leptocorisa oratorius (F.)		Rice bug
Leptocorisa tagalica Ahmad	*Leptocorisa geniculata* China	Rice bug
Lissorhoptrus oryzophilus Kush.		Rice water weevil
Maliarpha separatella Rag.		White stem borer

Pest	Some synonyms or misidentifications	Common name
Mythimna separata (Wlk.)	*Leucania separata* (Wlk.) *Pseudaletia separata* (Wlk.)	Rice armyworm, ear-cutting caterpillar
Mythimna unipuncta (Haw.)	*Cirphis unipuncta* (Haw.) *Pseudaletia unipuncta* (Haw.)	Common armyworm, ear-cutting caterpillar
Naranga aenescens Moore		Green rice caterpillar
Nephotettix cincticeps (Uhl.)		Green rice leafhopper
Nephotettix nigropictus (Stål)	*Nephotettix apicalis* (Motsch.) *Nephotettix yapicola* Linnavuori	Green rice leafhopper
Nephotettix parvus Ishihara & Kawase		
Nephotettix virescens (Dist.)	*Nephotettix impicticeps* Ishihara	Green rice leafhopper
Nezara viridula L.		Green vegetable bug, southern green stink bug
Nilaparvata lugens (Stål)		Brown planthopper
Nymphula depunctalis (Gn.)		Rice caseworm
Nymphula nympheata (L.)		China mark moth
Nymphula vittalis Brem.		Smaller rice caseworm
Oebalus ornata (Sailer)		Rice stink bug
Oebalus poecilus (Dall.)		
Oebalus pugnax (F.)		Rice stink bug
Orseolia oryzae (Wood-Mason)	*Pachydiplosis oryzae* (Wood-Mason)	Rice stem gall midge, maleng bug
Oryzaephilus mercator (Fauv.)	*Silvanus mercator* (Fauv.)	Merchant grain beetle
Oryzaephilus surinamensis (L.)	*Silvanus surinamensis* (L.)	Saw-toothed grain beetle
Oulema oryzae (Kuway.)	*Lema oryzae* (Kuway.)	Rice leaf beetle
Parnara guttata (Bremer & Grey)		
Pelopidas mathias (F.)		Rice skipper
Planococcoides lingnani (Ferris)	*Planococcus lingnani* Ferris	
Recilia dorsalis (Motsch.)	*Inazuma dorsalis* (Motsch.) *Deltocephalus dorsalis* (Motsch.)	Zig-zagged leafhopper
Rhopalosiphum padi (L.)	*Rhopalosiphum prunifoliae* (Fitch)	Oat bird, cherry aphis, apple grain aphis
Rhopalosiphum rufiabdominalis Sasaki	*Rhopalosiphum splendens* (Theob.)	Rice root aphid, red rice root aphid
Rhyzopertha dominica (F.)	*Pinoderus dominica* (F.)	Lesser grain borer
Rupela albinella (Cram.)	*Scirpophaga albinella* (Cram.)	South American white borer
Schizaphis graminum (Rond.)	*Toxoptera graminum* (Rond.)	

Pest	Some synonyms or misidentifications	Common name
Scotinophara coarctata (Thnb.)		Black paddy bug, Malayan black rice bug
Scotinophara lurida Burm.		Japanese black rice bug
Sesamia calamistis (Hmps.)		Mauritius pink borer of sugarcane
Sesamia inferens (Wlk.)		Pink borer
Seselia pusilla Gerst.		
Sipha glyceriae (Kalt.)	*Sipha schoutedeni* Del G.	
Sitophilus oryzae (L.)	*Calandra oryzae* (L.) *Sitophilus sasakii* (Tak.)	Rice weevil Small rice weevil
Sitophilus zeamais (Motsch.)		Maize weevil, large rice weevil
Sitotroga cerealella (Ol.)		Angoumois grain moth
Sogatella furcifera (Horv.)		White-backed planthopper
Sogatodes cubanus (Crwf.)	*Sogata cubana* (Crwf.)	Rice delphacid
Sogatodes orizicola (Muir)	*Sogata orizicola* Muir	Rice delphacid
Spodoptera exempta (Wlk.)	*Laphygma exempta* Wlk.	Armyworm
Spodoptera exigua (Hbst.)		Armyworm
Spodoptera frugiperda (J.E. Smith)	*Laphygma frugiperda* (J.E. Smith)	Fall armyworm
Spodoptera litura (F.)	*Prodenia litura* (F.)	Rice cutworm, common cutworm
Spodoptera mauritia (Boisd.)		Rice armyworm, rice swarming caterpillar
Stenocoris southwoodi Ahmad,	*Leptocorisa apicalis* Stål	
Telicota augias (L.)		Rice skipper
Tribolium castaneum (Hbst.)		Rust red flour beetle
Trichispa sericea (Guer)		Rice beetle
Tryporyza incertulas (Wlk.)	*Schoenobius incertulas* (Wlk.)	Yellow stem borer, paddy stem borer
Tryporyza innotata (Wlk.)	*Scirpophaga innotata* (Wlk.)	White rice borer, white stem borer
Unkanodes albifascia (Mats.)	*Delphacodes albifascia* (Mats.) *Ribautodelphax albifascia*	

MITE

Oligonychus oryzae (Hirst)		Paddy mite

CHECK LIST OF NEMATODES

Species	Some synonyms or misidentifications	Common name
Aphelenchoides besseyi Christie	*Aphelenchoides oryzae* Yokoo	Rice leaf or white tip nematode
Ditylenchus angustus (Butler) Filipjev		Rice stem nematode
Meloidogyne incognita (Kofoid & White) Chitwood	*M. incognita* var. *acrita* Chitwood	Root-knot nematodes
Meloidogyne javanica (Treub) Chitwood		
Meloidogyne graminicola Golden & Birchfield		
Heterodera oryzae Luc et Briz.		Rice cyst nematode
Heterodera sacchari Luc & Merny		Sugarcane cyst nematode
Hirschmanniella oryzae (Soltwedel) Luc & Goodey	*Radopholus oryzae* Thorne	Rice root or burrowing root nematode
Hirschmanniella spinicaudata (Sch. Stek.) Luc & Goodey	*Radopholus lavabri* Luc	
Hirschmanniella mucronata (Das) Luc & Goodey	*Radopholus mucronatus* Das	
Hirschmanniella imamuri Sher		
Macroposthonia onoensis (Luc) De Grisse	*Criconemoides onoense* Luc	Ring nematode
Tylenchorhynchus martini		Rice stunt nematode

PESTICIDE CHECK LIST

The inclusion of a trade name does not indicate approval of that product.

A = antibiotic, F = fungicide, H = herbicide, I = insecticide, N = nematicide

Common name	Trade names
acephate I	Orthene, Ortran
acephate + carbaryl I	Ortran-Nac fine granule F
alachlor H	Lasso
allyxycarb, APC (Japan) I	Hydrol
anilazine F	triazine, Triazine
azinphos-methyl I	Cotnion-methyl, Gusathion
BAS-3271 F F	N-cyclohexyl-2,5-dimethylfuran-3-carboxylamide
benomyl F	Benlate
benomyl + thiram F	20%a.i. + 20%a.i. = Benlate T
bentazon H	Basagran
benthiocarb H	Saturn, Bolero
γ-BHC (HCH, lindane) I	Gammexane, Dol, Sang-gamma
BHC + BPMC I	Gamma Bassa
BHC + carbaryl I	Sevidol G
BHC + isoprocarb I	Gamma-Hytox, Gamma Mipcin
BHC + MTMC I	Dolmix
blasticidin-S A	Bla-S
BPMC I	Bassa, Osbac
bromophos I	Nexion
bufencarb I	Bux granules
butachlor H	Machete
butralin H	Amex
captafol F	Difolatan, Folcid
captan F	Captan, Orthocide
carbaryl I	Denapon, NAC, Sevin
carbofuran I	Furadan
carbophenothion I	Garrathion, Trithion
cartap I	Padan
cartap + BPMC I	Padan Bassa granule 4
cellocidin A	Cellomate
chloramphenicol A	Shirahagen C
chlordimeform I	Galecron, Spanone
chlorfenvinphos, CVP (Japan) I	Birlane, Sapecron, Supona
chlormethoxynil H	X-52
chlornitrofen, CNP (Japan) H	Mo, CNP
chloropicrin I	Chloropicrin
chlorothalonil F	Daconil 1787
chlorpyrifos I	Dursban, Lorsban
cyanofenphos, CYP (Japan) I	Surecide
cypendazole F	Folcidin
cyperquat H	still under development
2,4-D H	numerous trade names
dalapon H	Dowpon, Basfapon, Gramevin
DBCP N	Fumazone, Nemagon, Nemanex, Nemaset
DCMP I	Fujithion

Common name	Trade names
D-D N	D-D
DDT I	numerous trade names
Demeton-methyl I	Metasystox
Demeton-S-methyl I	Metasystox (i), Metasystox 55
diazinon I	Basudin, Diazinon, Neocidal
diazinon + BPMC I	Bassazinon fine granules
diazinon + MTMC I	Tsumazinon fine granules
dichlorvos I	Dedevap, Nogos, Nuvan, Vapona
dicrotophos I	Bidrin, Carbicron, Ektafos
dimethoate I	Perfekthion, Rogor, Roxion, Tri Roxion
dinitramine H	Cobex
disulfoton I	Disyston, Ekatin TD
dithianon F	Delan
DSP N	Kaya-ace
EDB N	Dowfume W-85, Nemafume
edifenphos F	Hinosan
endosulfan I	Thiodan
EPN I	EPN
EPTC H	Eptam
ethyl DDD I	Perthane
fenitrothion I	Accothion, Agrothion, Sumithion, Folithion
fenitrothion + BPMC I	Sumibassa
fenitrothion + cyanofenphos I	Watathion
fenitrothion + malathion I	Ambithion
fenitrothion + MPMC I	Sumiace, Sumibal
fenitrothion + MTMC I	Tsumasumi fine granules
fenoprop (2,4,5-TP) H	Silvex
fensulfothion I	Dasanit, DMSP, Terracur-P
fenthion I	Baycid, Baytex, Lebaycid
fentiazon F	Celdion
fentin acetate FH	Brestan
ferric methane arsonate F	Neo-Asozin
fluorodifen H	Preforan
fluoronitrofen, CFNP (Japan) H	Mo-500
formothion I	Aflix, Anthio
formothion + BPMC I	Anthio-Bassa granule
formothion + disulfoton I	Anthio + Ekatin TD
glyphosate H	Roundup
hymexazol F	hydroxyisoxazole, Tachigaren
IBP F	Kitazin P
IKF-214 F	(not published)
Isoprocarb, MIPC (Japan) I	Etrofolan, Hytox, Mipcin
isothioate I	Hosdon
kasugamycin A	Kasumin
leptophos MBCP (Japan) I	Abar, Phosvel
malathion I	Cythion, and numerous others
mancozeb F	Dithane M-45
maneb F	Manzate, Dithane M-22
MCPA H	numerous trade names
mecarbam I	Murfotox, Pestan
mephosfolan I	Cytrolane
methomyl I	Lannate

Common name	Trade names
methyl arsenic sulphide F	MAS, Asozin
methyl bromide IFN (fumigant)	Kayafume, Dowfume, and numerous others
molinate H	Ordram
monocrotophos I	Azodrin, Nuvacron
MPMC I	Meobal
MTMC I	Tsumacide
MTMC + chlordimeform I	Tsumaspanon
naled I	Hibrom, Dibrom
nickel dimethyldithiocarbamate F	Sankel
nitrofen H	TOK, NIP
oxadiazon H	Ronstar
oxydemeton-methyl I	Metasystox R
paraquat H	Gramoxone
parathion-methyl I	Folidol-M
pentachlorobenzyl alcohol F	Blastin
perfluidone H	Destun
phenazine oxide F	Phenazine
phenthoate, PAP (Japan) I	Cidial, Elsan, Papthion
phorate I	Thimet
phosmet, PMP (Japan) I	Appa, Imidan, Prolate
phosphamidon I	Dimecron
piperophos + dimethametryn H	Avirosan
pirimiphos-ethyl I	Primicid
pirimiphos-methyl I	Actellic
polyoxins A	Polyoxin A
propanil, DCPA (Japan) H	STAM, Surcopur
propaphos I	Kayaphos
propoxur, PHC (Japan) I	Arprocarb, Baygon, Unden, Suncide
pyridaphenthion I	Ofunak
pyridaphenthion + MTMC I	Ofunak M fine granule
quintozene, PCNB (Japan) F	Brassicol
REE N	Sassen
salithion I	Salithion
SD 8280 I	2-chloro-1-(2,4-dichlorophenyl) vinyl dimethyl phosphate (Rangado)
2,4,5-T H	numerous trade names
TCE-styrene H	Tavron
tetrachlorvinphos I	Gardona
TF-128 F, TF-130 F	(thiadiazole compounds)
thionazin N	Nemafos
thiophanate-methyl F	Topsin-M, Cercobin
thiram F	Benlate T (20%a.i. benomyl + 20%a.i. thiram), Arasan, Thiuram
triazophos I	Hostathion
trichlorphon, DEP (Japan) I	Dipterex
trifluralin H	Treflan
validamycin A	Validacin
vamidothion I	Kilvar
XMC I	Cosban, Macbal

INDEX

F

fall armyworm see *Spodoptera frugiperda*
false smut see *Ustilaginoidea virens*
fenitrothion
    use in rice storage   250, 251
    use to control *Aphelenchoides*
                  besseyi   97
                  Baliothrips biformis   157
                  Eusarcoris
                    inconspicuus   156
                  hispas   182
                  Hydrellia griseola   177
                  Mythimna spp.   173
                  Naranga aenescens   174
                  Nephotettix spp.   134
                  Orseolia oryzae   196
                  Oulema oryzae   183
                  planthoppers   140
                  Scotinophara spp.   153
                  Sogatodes spp.   146
                  Spodoptera spp.   172
                  stem borers   122
                  virus vectors   80, 86
fenoprop
    use in weed control   30, 33, 35
fensulfothion
    use to control *Aphelenchoides*
                  besseyi   97
                  Macroposthonia
                    onoensis   104
                  Nephotettix spp.   135
                  stem borers   123, 126
fenthion
    use in bird control   214
    use to control *Agromyza oryzae*   174
                  Aphelenchoides
                    besseyi   97
                  Baliothrips biformis   157
                  Chlorops oryzae   194
                  grasshoppers   133
                  hispas   182
                  Hydrellia griseola   177
                  Nephotettix spp.   135
                  Nezara viridula   155
                  Nymphula spp.   163
                  Oulema oryzae   183
                  planthoppers   141
                  rice bugs   149
                  Scotinophara spp.   153
                  stem borers   123
fentiazon
    use to control *Xanthomonas oryzae*   75

fentin acetate
    use to control algae   33
                  snails   108
ferns   21
ferric methane arsonate
    use to control *Corticium sasakii*   60
fertilizer   90—91
    application   28, 56
    carrier for insecticides   265
Fimbristylis littoralis   21
Fimbristylis miliacea   15
floating rice   12
    weed control   37
flooding
    use to control insect pests 153, 172, 198
                  nematodes   101
fluoroacetamide
    use to control rodents   230
fluorodifen
    use in weed control   30, 34
fluoronitrofen
    use in weed control   32
foot rot see *Gibberella fujikuroi*
formothion
    use to control *Cnaphalocrocis*
                  medinalis   160
                  Nephotettix spp.   135
                  stem borers   123
Foudia madagascariensis   207
Fuirena sp.
    alternative host of *Macroposthonia*
                  onoensis   104
fulvous duck see *Dendrocygna fulva*
fumigants for stored rice   252
fungal diseases   43—72
Fusarium moniliforme see *Gibberella*
        fujikuroi
Fusarium nivale   53
Fusarium spp.   63

G

galls
    on leaves   194—196
    roots   99—101
garganey see *Anas querquedula*
Geococcus oryzae   189
giallume   81
Gibberella fujikuroi   51, 64, 239
Glyceria acutiflora
    alternative host of yellow dwarf   89
glyphosate
    use in weed control   28, 30, 35, 37, 38